Hyperbaric Oxygen Therapy: Principles and Applications

Manoj Gupta • Indumathi Somasundaram

Hyperbaric Oxygen Therapy: Principles and Applications

Springer

Manoj Gupta
Sailee Hospital and Diagnostic Centre
Prana Hyperbaric Oxygen Therapy Centre
Mumbai, Maharashtra, India

Indumathi Somasundaram
Department of Biotechnology Engineering
Kolhapur Institute of Technology's College of Engineering
Kolhapur, Maharashtra, India

ISBN 978-981-99-7280-7 ISBN 978-981-99-7278-4 (eBook)
https://doi.org/10.1007/978-981-99-7278-4

This Springer imprint is published by the registered company Springer Nature Singapore Pte Ltd.
The registered company address is: 152 Beach Road, #21-01/04 Gateway East, Singapore 189721, Singapore

Paper in this product is recyclable

Contents

Abbreviations

ADH	Anti-diuretic hormone
AGE	Arterial gas embolism
AHA	American Hyperbaric Association
AMA	American Medical Association
ANF	Atrial natriuretic factor
ATA	Atmospheric absolute
bFGF	Basic fibroblast growth factor
CNS	Central nervous system
CRAO	Central retinal artery occlusion
DCS	Decompression sickness
ECHM	European Committee for Hyperbaric Medicine
ECM	Extracellular matrix
EFA	Euglobulin fibrinolytic activity
eNOS	Endothelial NOS
EPC	Endothelial precursor cells
ETC	Electron transport chain
$FADH_2$	Flavin adenine dinucleotide hydrogen
FHN	Femoral head necrosis
GABA	Gamma amino butyric acid
GPx	Glutathione peroxidase
GRx	Glutathione reductase
HBOT	Hyperbaric oxygen therapy
HIF	Hypoxia induced factors
HPNS	High pressure neurological syndrome
ICA	Intracranial abscess
ICAM1	Intercellular adhesion molecule 1
IGF-1	Insulin-like growth factor 1
IL	Interleukins
ISSHL	Idiopathic sudden sensorineural hearing loss
LPS	Lipopolysaccharide
MEB	Middle ear barotrauma
MOE	Malignant otitis externa
NADH	Nicotinamide adenine dinucleotide hydrogen
NMDA	N-methyl-D-aspartate

pAO_2	Alveolar partial pressure of oxygen
PBMCs	Peripheral blood mononuclear cells
PBT	Pulmonary barotrauma
pCO_2	Partial pressure of CO_2
PDGF	Platelet derived growth factor
ROS	Reactive oxygen species
SNP	Single nucleotide polymorphism
SOD	Superoxide dismutase
TCOM	Transcutaneous oximetry
$TcPO_2$	Transcutaneous oxygen tension
TGF-β	Transforming growth factor-β
TLR2	Toll-like receptors 2
t-PA	Tissue type plasminogen activator
TRPV	Transient receptor potential vanilloid
UHMS	Undersea and Hyperbaric Medical Society
UPTD	Unit pulmonary toxic dose
VCAM1	Vascular cell adhesion molecule 1
VEGF	Vascular endothelial growth factor
VGE	Venous gas embolism
VOC	Vaso-occlusive crisis

History and Development of Hyperbaric Medicine

1

The origin of hyperbaric medicine is derived from the earlier observations of pressure implications and physiological effects of deep-sea diving. In order to extend the diving time and depth, specialized chambers or barrels had been designed [1]. The first such basic diving apparatus has been documented to be used by Alexander the Great in 320 B.C. Leonardo Da Vinci made sketches of similar diving equipment which were never materialized. In 1620, a Dutch inventor Cornelius Drebbel devised a diving bell which can be pressurized up to 1 atmospheric absolute (ATA). Later, in 1691, Edmund Halley used weighted barrels in the diving bell in order to replenish air supply during diving [2]. With the advent of modern technology in the last two centuries, more efficient air compressed chambers and diving suits were made. However, the exposure to high pressure during deep-sea diving results in a clinical condition called as decompression sickness and hyperbaric medicine is found to be effective in its treatment [3]. The development of hyperbaric medicine till date occurred in the course of nearly four centuries. Hyperbaric medicine or hyperbaric oxygen therapy (HBOT) was first used in the treatment of tuberculosis and presently has been approved and found to be effective in the treatment of 14 different clinical conditions [1, 4]. The earlier hyperbaric applications and equipment were designed based on the general observation of physiological effects of pressure on human body and had less scientific rationale behind the therapeutic practices. The advancement in understanding of physical laws governing gases and functioning of respiratory system with respect to oxygen uptake and distribution enabled physicians to effectively utilize the oxygen pressure effects for therapeutic purposes (Fig. 1.1).

M. Gupta, I. Somasundaram, *Hyperbaric Oxygen Therapy: Principles and Applications*, https://doi.org/10.1007/978-981-99-7278-4_1

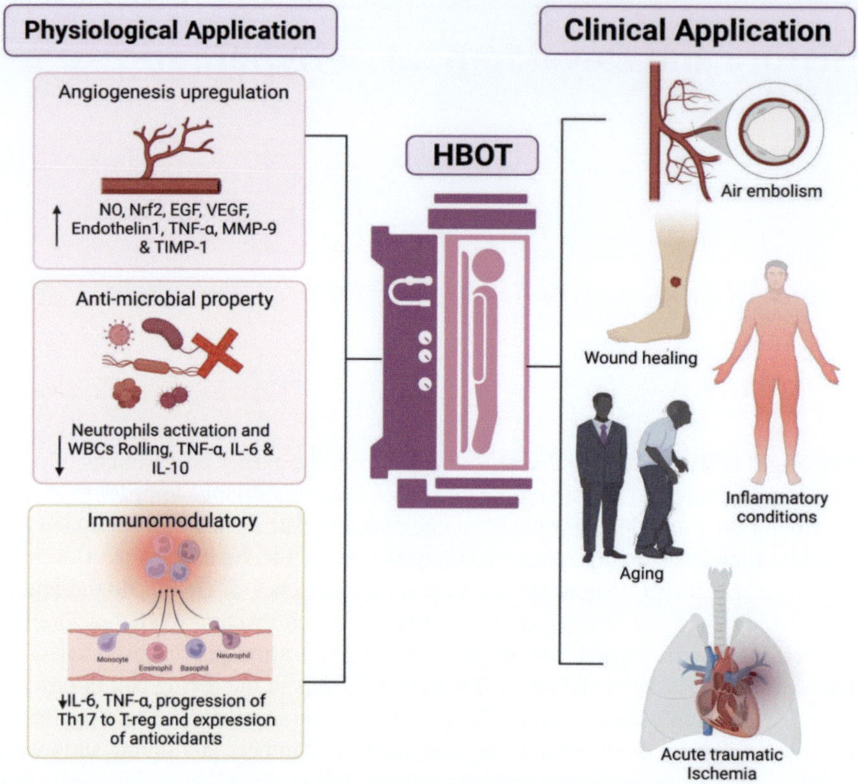

Fig. 1.1 HBOT has been used in a wide array of clinical applications by virtue of its ability to enhance multiple physiological functions like angiogenesis, protection against infection, and augmentation of immune functions

1.1 Brief History of Hyperbaric Medicine

In 1662, a British clergyman and physician Nathaniel Henshaw first utilized atmospheric pressure for therapeutic purpose. Hyperbaric oxygen therapy has been developed and utilized successfully till date for a large number of medical purposes (Fig. 1.2). Henshaw first created a closed chamber termed as "domicilium" attached with a pair of organ bellows. The pressure inside the chamber could be regulated by a series of valves connected to the bellows. Henshaw's design allowed for modest changes in ambient pressure and is utilized for the treatment of chronic ailments related to respiratory and digestive systems. The design had many major drawbacks such as retention of metabolic waste gases being an unventilated system and limited use in therapy due to the compression of ambient atmospheric gases instead of purified oxygen [1, 5]. Henshaw's work did not provide any scientific rationale for the effects of hyperbaric environment on clinical conditions but it led to the foundation for the development of hyperbaric medicine.

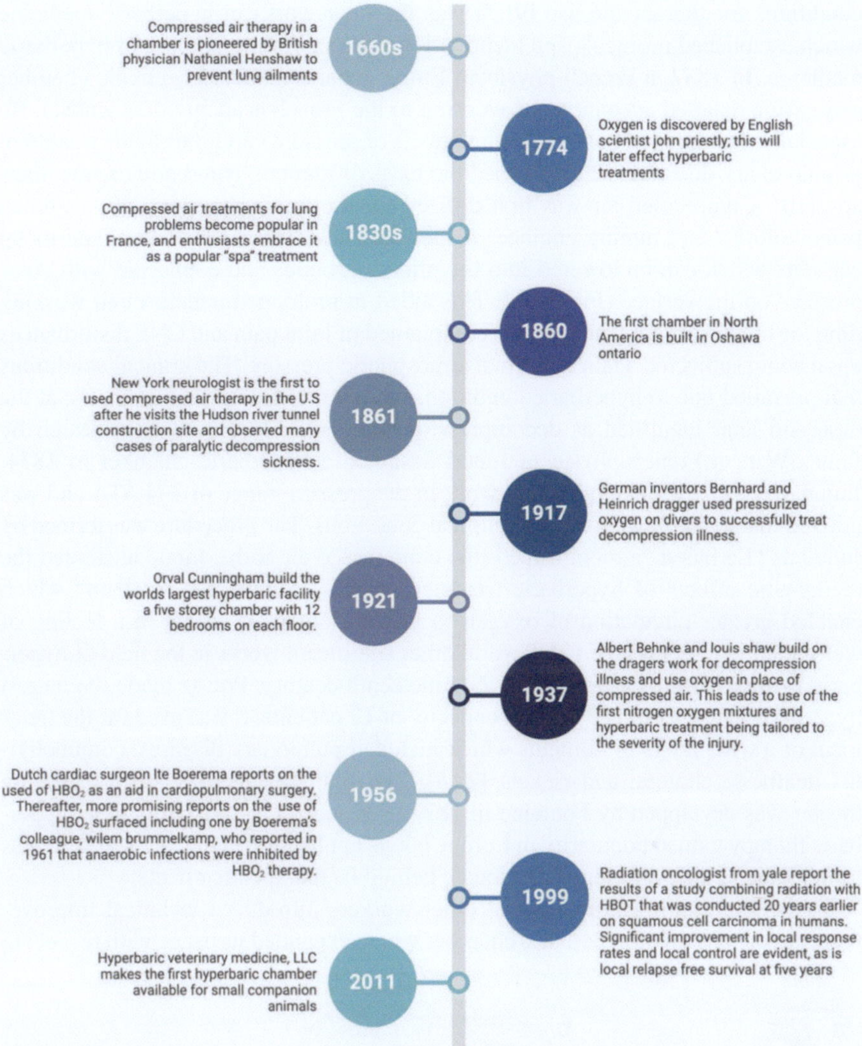

Fig. 1.2 Historical timeline for the development of hyperbaric medicine

Discovery of oxygen by John Priestly in 1775 had a great impact in the development of hyperbaric treatment. There were significant insights given by Lavoisier and Seguin in 1789 regarding the toxic effects of concentrated oxygen [6, 7]. This resulted in less inclination toward utilizing compressed oxygen for therapeutic purpose and most workers on hyperbaric therapy suggested application of normobaric conditions instead of hyperbaric conditions. Beddoes and Watt wrote the first book on oxygen therapy in 1796 where normobaric conditions were emphasized for treatment [8]. Similarly, Paul Bert, who is considered as the father of pressure physiology, explained the scientific basis of oxygen toxicity and advocated normobaric conditions for therapeutic use [9]. It was the later works in hyperbaric medicine which established protocols and identified limitations for the safe use of hyperbaric medicine. In 1832, a French physician Emile Tabarie made a pneumatic chamber and gave a detailed account of its working to the French academy of scientists. He especially documented the therapeutic effects of pressure variations in the treatment of pulmonary diseases, which formed the basis of modern hyperbaric oxygen therapy [10]. Compressed air was first utilized for the mining purposes by a French paleontologist and mining engineer named Jacques Triger. Triger designed metal caissons which can be lowered into the mining grounds and connected with compressors on the surface (Fig. 1.3a). This aided in prolonged underground working time for mine workers. The workers complained of joint pain and CNS disturbances upon being subjected again to normal atmospheric pressure. The clinical conditions that prevailed due to hyperbaric conditions were termed as caisson's disease at the time and later identified as decompression sickness [11]. Based on a design by James Watt, a French physician Junod designed a hyperbaric chamber in 1834. Junod's hyperbaric chamber could work in the pressure range of 2–4 ATA and was utilized in the treatment of various clinical conditions. The procedure was termed by Junod as "Le baind "aircomprime" (the compressed air bath). Junod attributed the therapeutic effects of hyperbaric treatment to the "compressed air bath" which enabled greater circulation of oxygen to internal organs resulting in a feeling of well-being in patients [10, 12]. Several other significant works in the field of hyperbaric medicine were carried out in the nineteenth century. Pravaz made the largest hyperbaric chamber in 1837 with a capacity of 12 patients. It was used for the treatment of a wide range of ailments which included pulmonary diseases, conjunctivitis, deafness, cholera, and rickets [2, 5]. The first mobile hyperbaric operating theater was developed by Fontaine in 1877 (Fig. 1.3b) [2, 5]. By this time, hyperbaric therapy gained popularity in Europe but its application in the medical field was limited due to lack of scientific rationale behind its therapeutic effects.

In the twentieth century, several other workers introduced technical improvements in the design of hyperbaric chamber which expanded its usage with respect to

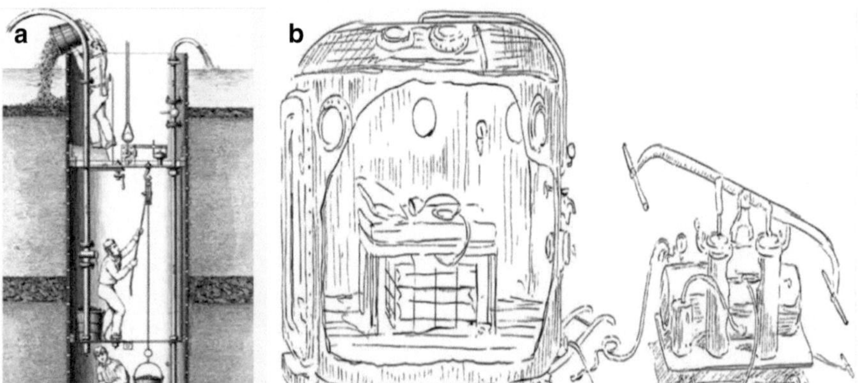

Fig. 1.3 (**a**) Triger's Caisson. (**b**) Fontaine's mobile operating room

efficiency and clinical applications. One such significant contribution in this field was given by an American physician named Orval J Cunningham. In 1921, during the final days of World War I, Cunningham made a hyperbaric chamber for the treating the victims of Spanish influenza epidemic. He observed that the mortality rates from the disease were higher in the areas of higher altitudes and suggested that barometric factor is critical for the etiology of disease. Cunningham claimed to be able to successfully treat patients who were under cyanotic and comatose conditions. Cunningham hyperbaric facility gained popularity and he continued to apply hyperbaric treatment for diseases such as cancer, diabetes mellitus, hypertension, and syphilis. He stated that the therapeutic effects of hyperbaric treatment arise from the hyperoxygenating conditions which aid in curbing down the anaerobic infections responsible for most of the diseases. It was during this time that the first major fire incident associated with hyperbaric chamber occurred due to manual error. Fortunately, all the patients were evacuated in time and there was no loss of life. Another tragic incident occurred shortly afterwards leading to the death of all his patients under treatment when there was a mechanical failure resulting in complete loss of compression in the chamber [13]. Such incidents raised questions about the safety of using hyperbaric treatment and prompted hyperbaric practitioners to follow strict safety protocols. Cunningham continued his work on hyperbaric medicine and constructed world's largest hyperbaric facility in Cleveland, Ohio, in 1928. It consisted of five stories with five rooms each, all dedicated to hyperbaric treatment. Cunningham's work gained attention of Bureau of investigation of American Medical Association (AMA), which demanded scientific evidences for his claims regarding the hyperbaric oxygen therapy. Cunningham failed to provide any major proof to validate his findings before AMA resulting in the loss of credibility in his work with hyperbaric medicine [14].

One of the notable works in the field of hyperbaric medicine involved the use of oxygen under hyperbaric condition and examining the sufficiency of dissolved oxygen in animals (piglets) after replacing the blood with plasma. Boerema et al., in 1959, carried out experiments on piglets by replacing their blood with plasma or rheomacrodex such that the hemoglobin (Hb) levels were lowered up to 0.4%. The animals were operated in a pressure chamber at 3 ATA pressure and were supplied with pure oxygen. The only source of oxygen transport in the absence of Hb was plasma dissolved oxygen and the animals were restored with blood at the end of the procedure. The animals recovered successfully and showed no sign of disturbances over the next few weeks [15].

1.2 HBOT and Radiosensitization

HBOT results in the enhanced dissolution of oxygen in the blood plasma. The physiological effects of HBOT are manifested in the form of increased vasoconstriction reduction of edema, phagocytosis, anti-inflammation, and increased tissue oxygenation [16, 17]. The virtue of increased tissue oxygenation by administration of HBOT has been utilized in cancer treatment by radiation therapy. There are usually two types of clinical practice for the application of HBOT in conjunction with radiotherapy. First approach involves administration of HBOT simultaneously or prior to radiotherapy in order to sensitize the tumor cells to radiation treatment, a process known as radiosensitization. Usually, there is radiation-induced damage to the normal healthy tissues in the form of lesions. Second approach of using HBOT involves treatment and reduction of radiation-induced damage to normal tissues [18, 19].

Tumor cells have a hypoxic environment which makes them resistant to the effects of radiotherapy [20]. Administration of HBOT during or prior to radiotherapy leads to increased levels of oxygen in the tumor cells and sensitizing them to radiotherapy. Radiosensitization of tumor cells by using HBOT was first used and reported by Churchill Davidson in 1968 [21]. There are certain other techniques for the purpose of increasing oxygen pressure in tumor cells such as administration of red blood cells or administration of radiation sensitizing agents like nitroimidazoles [22–24]. HBOT promotes enhanced fibroblastic activity and collagen production leading to extracellular matrix formation and neovascularization [25, 26]. The additional effects of HBOT in curbing the negative effects of radiotherapy make it a preferential choice of method for radiosensitization (Fig. 1.4).

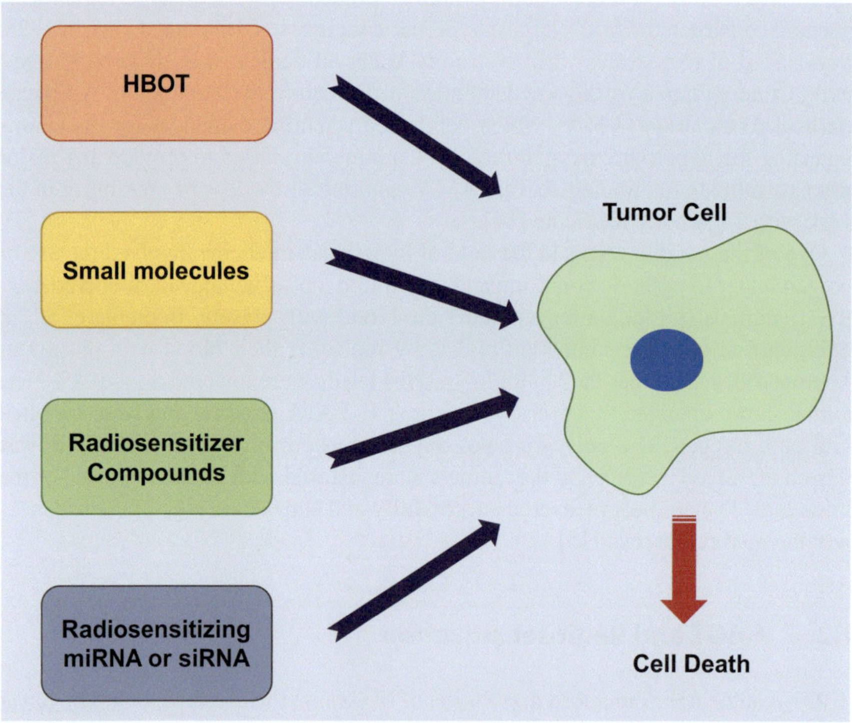

Fig. 1.4 Radiosensitization. Various small molecules, radiosensitizer compounds, and radiosensitizing miRNA or siRNA are utilized to enhance the effect of radiation therapy against tumor cells. HBOT has been found effective for the same purpose by augmenting the production of free radicals which aid in tumor cell killing in radiation therapy

1.3 Future Perspectives of HBOT in Therapeutics

Currently, with a much better understanding of the underlying principles of hyperbaric medicine, it has gained popularity as well as credibility equally among physicians and patients. Also, there are clear scientific evidences in the treatment of a wide range of clinical conditions and a better understanding of scientific rationale behind the effects of HBOT. Apart from its applications as a conjunctive therapy in a large number of ailments, there are various potential benefits of HBOT which are under study. Its effects in wound healing and regeneration are attributed mainly to the physiological effects like vascularization, anti-inflammation, etc. But recent studies have indicated its role in genetic and epigenetic modulation of several crucial genes involved in healing and regeneration. Thus, HBOT shows promise in the field of regenerative medicine and anti-aging therapies as well. However, more research and scientific evidences are required to establish HBOT as a regenerative therapy.

References

1. Jain KK. The history of hyperbaric medicine. In: Textbook of hyperbaric medicine. Cham: Springer; 2017. p. 3–9.
2. Wattel F. A history of hyperbaric medicine. In: Handbook on hyperbaric medicine. Springer; 2006. p. 1–11.
3. Shim SS, Patterson F, Kendall M. Hyperbaric chamber and decompression sickness: an experimental study. Can Med Assoc J. 1967;97(21):1263.
4. Mathieu D, Marroni A, Kot J. Tenth European consensus conference on hyperbaric medicine: recommendations for accepted and non-accepted clinical indications and practice of hyperbaric oxygen treatment. Diving Hyperb Med. 2017;47(1):24.
5. Morrison DS and Kirkby RD. Hyperbaric medicine: a brief history. 5. http://www.quackwatch.org/01Quackery RelatedTopics/HBOT/hm01.html. Accesed July 2001. 2001.
6. Bean JW. Effects of oxygen at increased pressure. Physiol Rev. 1945;25(1):1–147.
7. Hancock JT. A brief history of oxygen: 250 years on. Oxygen. 2022;2(1):31–9.
8. Cartwright FF. The association of Thomas Beddoes, MD with James Watt, FR S. Notes Rec R Soc Lond. 1967;22(1):131–43.
9. Dejours P, Dejours S. The effects of barometric pressure according to Paul Bert: the question today. Int J Sports Med. 1992;13(S 1):S1–5.
10. Stewart J. Exploring the history of hyperbaric chambers, atmospheric diving suits and manned submersibles: the scientists and machinery. Bloomington: Xlibris Corporation; 2011.
11. Glossop R. The invention and early use of compressed air to exclude water from shafts and tunnels during construction. Geotechnique. 1976;26(2):253–80.
12. Doran L, Riordan N. Hyperbaric oxygen treatment.
13. Alvarez WC. Hyperbaric oxygenation: Vol 117, article 2, pp 647-890. Arch Intern Med. 1966;118(1):91–2.
14. Krishnamurti C. Historical aspects of hyperbaric physiology and medicine. In: Respiratory physiology. London: IntechOpen; 2020. p. 1–18.
15. Hart GB. Exceptional blood loss anemia: treatment with hyperbaric oxygen. JAMA. 1974;228(8):1028–9.
16. Braswell, C. and D.T. Crowe, Hyperbaric oxygen therapy. Compendium: Continuing Education for Veterinarians, 2012. 34(3): p. E1-E6.

17. Neuman TS, Thom SR. Physiology and medicine of hyperbaric oxygen therapy. Amsterdam: Elsevier Health Sciences; 2008.
18. Fernández E, et al. Hyperbaric oxygen and radiation therapy: a review. Clin Transl Oncol. 2021;23(6):1047–53.
19. Krajčovičová Z, Kulišiak VMJSP, Kašlíková K. Hyperbaric oxygen therapy as adjunctive strategy in radiotherapy. Univ Rev. 2018;12:1–4.
20. Vaupel P, Kelleher DK, Höckel M. Oxygenation status of malignant tumors: pathogenesis of hypoxia and significance for tumor therapy. In: Seminars in oncology, vol. 28. Amsterdam: Elsevier; 2001. p. 29.
21. Churchill-Davidson I. Therapeutic uses of hyperbaric oxygen. Ann R Coll Surg Engl. 1966;39(3):164.
22. Adams G, et al. Radiation sensitization and chemopotentiation: RSU 1069, a compound more efficient than misonidazole in vitro and in vivo. Br J Cancer. 1984;49(5):571–7.
23. Zayed S, et al. Red blood cell transfusion practices for patients with cervical cancer undergoing radiotherapy. JAMA Netw Open. 2021;4(4):e213531.
24. Yu M, et al. Oxygen carriers and cancer chemo-and radiotherapy sensitization: bench to bedside and back. Cancer Treat Rev. 2007;33(8):757–61.
25. Ferguson B, Hudson WR, Farmer JC Jr. Hyperbaric oxygen therapy for laryngeal radionecrosis. Ann Otol Rhinol Laryngol. 1987;96(1):1–6.
26. Spiegelberg L, et al. Hyperbaric oxygen therapy in the management of radiation-induced injury in the head and neck region: a review of the literature. J Oral Maxillofac Surg. 2010;68(8):1732–9.

Physiology of Oxygen Transport and Gas Laws

2

2.1 Composition of Air

The atmosphere consists of a mixture of 78.08% nitrogen, 21% oxygen, 0.04% carbon dioxide, and traces of remaining gases [1]. The pressure exerted by the atmospheric gases constitutes the atmospheric pressure which is measured to be 760 mm of mercury (mmHg). Each component of gas contributes to the pressure in proportion to its volume in the atmosphere, which can be calculated in terms of its partial pressure as:

$$\text{Partial pressure of gas} = (\text{absolute pressure}) \times (\text{proportion of total volume of gas}).$$

Thus, the partial pressure of oxygen can be calculated as:

$$760 \times (21/100) = 160\,\text{mmHg}.$$

Density of a gas is another important factor which is affected by the atmospheric pressure. Density is the measure of mass per unit volume, i.e. an object of high density will acquire less volume for a given mass as compared to the object of lower density. Air density depends upon a number of factors like temperature, pressure, and humidity [2]. All the atmospheric gases have different relative densities. The overall air density varies directly with pressure and inversely with temperature. Higher pressure results in reduction of intermolecular distance of gases, while increase in temperature has the opposite effect [3].

Water vapors are another important component of atmosphere which contributes to atmospheric pressure. The water vapor content in atmosphere is variable and depends on the temperature. Oxygen is a colorless, tasteless, and odorless gas existing as a diatomic molecule with molecular weight of 32. Oxygen consists of two unpaired electrons making it highly electronegative. It forms chemical bonds with nearly all the other elements to form oxides. Oxygen is essential for the existence of all aerobic life forms. Nitrogen is also a colorless, tasteless, and odorless gas and exists as a diatomic molecule. It is mostly a non-reactive gas and is insoluble in

M. Gupta, I. Somasundaram, *Hyperbaric Oxygen Therapy: Principles and Applications*, https://doi.org/10.1007/978-981-99-7278-4_2

water which leads to bubble formation in blood and tissues during decompression sickness. Carbon dioxide consists of one carbon atom bonded by each of two oxygen atoms by a double bond. It is a less reactive gas due to its full oxidized state and shows a high solubility in water. Nearly 1% of dissolved carbon dioxide reacts with water to form carbonic acid, which readily dissociates into bicarbonate and carbonate ions [4].

2.2 Gas Laws

The physical behavior of gases is governed by a set of laws known as gas laws and there are several factors which determine the properties of gas in a biological system such as solubility, temperature, and diffusion.

2.2.1 Henry's Law

This law was formulated by William Henry in 1803 which relates the solubility of a gas in liquid medium as a function of pressure and solubility coefficient of the gas. According to Henry's law:

$$\text{Concentration of dissolved gas} = \text{Pressure} \times \text{Solubility Coefficient}$$

Solubility coefficient is specific for every gas or liquid and is inversely proportional to the temperature. The solubility of respiratory gases in plasma or blood, i.e. O_2 and CO_2 can be calculated at physiological temperature by using Henry's law. Oxygen is dissolved at 0.024 ml O_2/ml, while CO_2 is dissolved at 0.5 ml CO_2/ml in plasma [5].

2.2.2 Dalton's Law

This law was given by John Dalton in 1801 which states that:

"The pressure exerted by a mixture of gases in a given volume is equal to the sum of pressure exerted by each gas if they alone occupied the same volume."

It may be represented as:

$$P_{Total} = P_1 + P_2 + P_3 \ldots P_n$$

Here, P_1,P_2, P_3.... P_n represent partial pressure of each individual gas in the mixture.

It can be stated that each gas contributes to the total pressure independently. Dalton's law can be used to calculate the partial pressure of individual gases in a gaseous mixture as a fraction of total pressure exerted. The partial pressure of the gas can be expressed as a product of total pressure of the gaseous mixture (P_{Total}) and the fraction of the gas (F_1) [6].

$$P_1 = P_{Total} \times F_1$$

The atmosphere comprises a mixture of different gases and the partial pressure of individual constituent gas varies with elevation in total pressure exerted. Thus, the gases may exhibit toxicity at an elevated pressure due to the virtue of their increased partial pressure. This consideration is important in designing the course of treatment in hyperbaric therapy.

2.2.3 Boyle's Law

Robert Boyle postulated this law which relates the pressure of a gas to its volume in a confined space. It states that:

"The pressure exerted by a given quantity of gas is inversely proportional to the volume of the gas at constant temperature."

$$\text{Thus, } p \times V = k(\text{Constant})$$

Specific amount of gas at a constant temperature in two different states designated by 1 and 2 can be related as:

$$P_1V_1 = P_2V_2$$

Boyle's law has a practical relevance in hyperbaric treatment as it relates the confined amount of gas in human body and hyperbaric equipment and governs the process of compression and decompression [5].

2.2.4 Amonton's Law/Gay-Lussac's Law

This law was given by Guillaume Amonton and published in detail by Thomas Grahamand, thus also known as Graham's law. It is also known as Gay-Lussac's law as he was the first to publish convincing evidence to show the relationship of temperature and pressure for a gas at a constant volume. It provides the relationship between pressure and temperature of a gas [7]. According to Amonton's law:

"At a constant volume of gas, the pressure of a given mass of gas varies directly with the absolute temperature of the gas."

In other words, the ratio of pressure and absolute temperature of a gas is constant at a given volume. Mathematically it can be expressed as:

$$\mathrm{p} / \mathrm{T} = \mathrm{K}(\text{Constant}) \text{ or } \mathrm{p}_1 / \mathrm{T}_1 = \mathrm{p}_2 / \mathrm{T}_2$$

It is important in considering the pressure changes during the compression in hyperbaric treatment as the temperature variations occur simultaneously.

2.2.5 Ideal Gas Law

Ideal gas law or universal gas law is a relationship between temperature, pressure, and volume of a gas and is a combination of Boyle's law and Amonton's law [8]. Mathematically it can be stated as:

$$\mathrm{p} \times \mathrm{V} / \mathrm{T} = \mathrm{Constant\ or\ p_1} \times \mathrm{V_1} / \mathrm{T_1} = \mathrm{p_2} \times \mathrm{V_2} / \mathrm{T_2}$$

2.2.6 Fick's Law of Gas Diffusion

Law of gas diffusion was derived by Adolf Fick in 1858. It relates the rate of gas diffusion across a membrane to the partial pressure of the gas [9]. As per Fick's law:

$$\text{Rate of diffusion} = K \times A \times \Delta P / D.$$

Here,

K = Constant dependent on temperature and gas type.
A = Surface area of membrane.
ΔP = Difference in partial pressure of gas on both sides of membrane.
D = Distance over which diffusion occurs (equivalent to membrane thickness).

Fick's law of diffusion is helpful in understanding the gaseous exchanges within the body, i.e. oxygen transport in lungs (from alveoli to blood capillaries) and in tissues (from blood to cells through plasma membrane) [5].

2.2.7 Adiabatic Compression of Gases and Joule–Thomson Effect

An adiabatic process is defined as a thermodynamic process in which there is no exchange of heat or matter between the system and the environment [10]. Thus, there is no change in the temperature of an adiabatic system. Joule–Thomson effect was postulated by James Prescott Joule and William Thomson, which defines the temperature changes during the adiabatic compression or decompression of a gas [11]. When an ideal gas is compressed adiabatically, i.e. without any external cooling, work is done by the gas molecules resulting in increase in the temperature of gas (Fig. 2.1). Conversely, during adiabatic expansion or decompression, there is a drop in temperature.

Since hyperbaric oxygen therapy involves the process of compression and decompression, the changes in temperature need to be carefully monitored and corrected during the whole process of treatment.

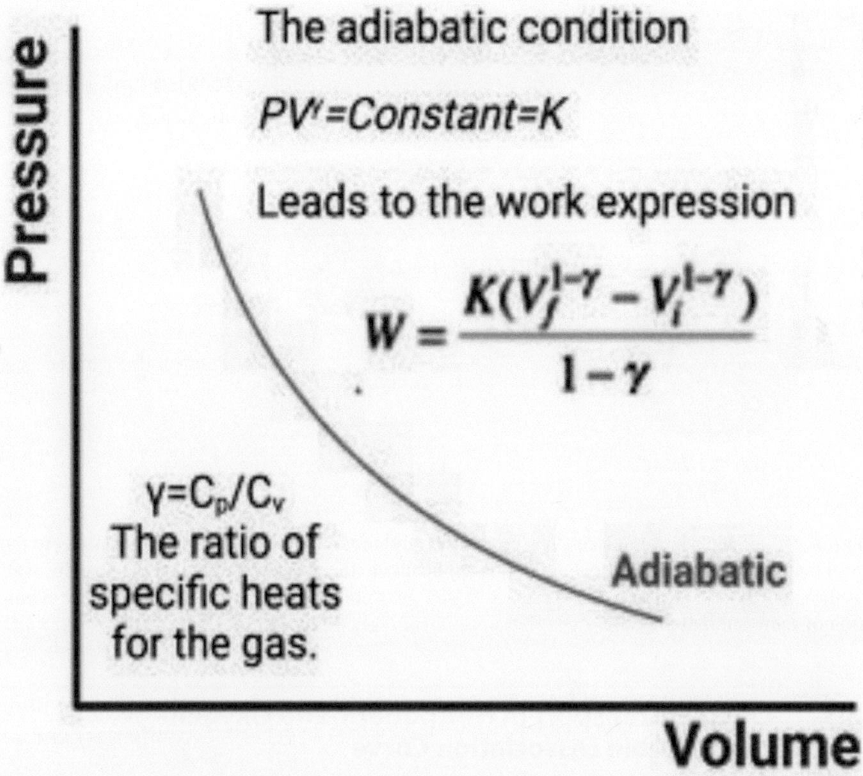

Fig. 2.1 Graphical representation for adiabatic compression of gases. Adiabatic process is defined as a process accompanied by neither loss nor gain of heat by the system. During adiabatic compression of gases, work is done by the gases resulting in rise in temperature. Conversely, during adiabatic expansion, there is a drop in temperature of the system

2.2.8 LaPlace's Law

Laplace's law describes the relation between internal pressure of a fluid and the wall tension and radius of a cylinder or sphere [12]. This law is applicable to the pressure exerted by blood on the blood vessels and chambers of heart (Fig. 2.2). The law states that:

"The larger the radius of the vessel, the larger will be the wall tension required to withstand a given internal fluid pressure."

Mathematically, Laplace's law can be expressed as:

$$\text{Tension}(T) = \text{Pressure}(P) \times \text{Radius}(R)/2$$

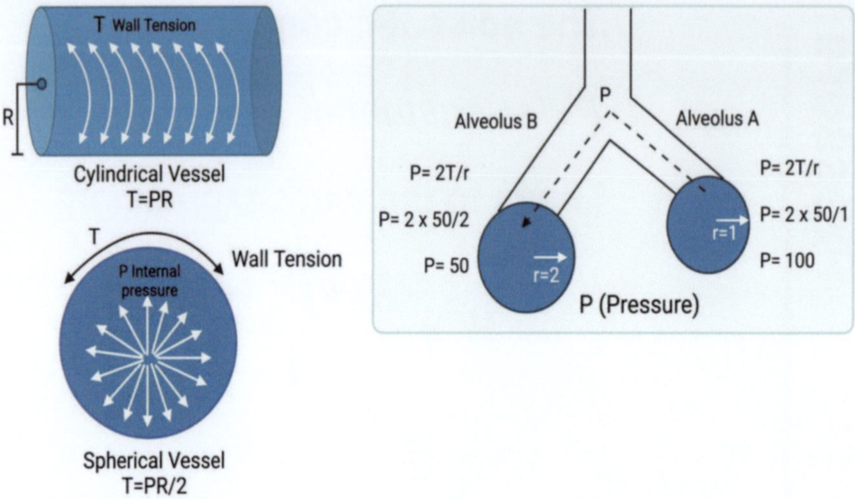

Fig. 2.2 LaPlace's law. Laplace's law relates the fluid pressure in a closed circular chamber to the wall tension and radius of the chamber. It is applicable to the pressure exerted on the alveolus wall as the small alveoli are likely to experience higher pressure than the larger alveoli under the condition of same internal pressure

2.3 Oxygen Transport in Normobaric and Oxygen–Hemoglobin Dissociation Curve

Oxygen is vital to aerobic life form and central to the process of cellular respiration. The process of respiration involves the intake of air through lungs, its diffusion to the pulmonary capillaries through alveoli, and transport of oxygen majorly in hemoglobin bound state to all the tissues for cellular respiration [13]. Carbon dioxide produced during cellular respiration is carried back to the alveolar space and exhaled through lungs. The rate of oxygen absorption depends on the alveolar ventilation or alveolar partial pressure of oxygen (pAO_2). Under normal conditions, the pAO_2 is maintained at 104 mmHg, while in case of moderate exercise, it can increase up to four folds the normal value. The partial pressure of O_2 in pulmonary capillaries is 40 mmHg which enables the diffusion of O_2 from alveoli into the pulmonary blood circulation. The partial pressure of CO_2 (pCO_2) in alveoli depends upon the rate of cellular respiration and is maintained at 40 mmHg under normal conditions [14, 15].

Around 97% of oxygen is transported from the lungs to different tissues in a bound state with oxygen carrying pigment hemoglobin, while 3% oxygen is transported in dissolved state in plasma. One molecule of Hb can bind with 4 oxygen molecules. The pO_2 in the tissues is maintained at 20–40 mmHg by gradual release of O_2 from Hb in the tissues [16, 17]. Under hyperbaric conditions, the rate of oxygen transport is only mildly affected as the Hb gets readily saturated under normobaric conditions due to cooperative binding, i.e. binding of one oxygen molecule facilitates the subsequent binding of next oxygen molecules [18]. Under this condition, around 97% of Hb is already saturated and it requires a very high pO_2 to saturate the remaining 3% Hb. Thus, increase in the concentration of inspired oxygen by increase in pressure does not elevate the rate of oxygen transport [19, 20]. The mechanism of oxygen binding and Hb saturation with oxygen can be explained by oxygen–hemoglobin dissociation curve or oxyhemoglobin curve. The oxygen–hemoglobin dissociation curve is a sigmoidal curve relating the partial pressure of oxygen to the percentage oxygen saturation of Hb. The binding affinity of oxygen to Hb is represented by P50, which is the partial pressure of oxygen at which half of the binding sites of Hb are saturated or in other words, Hb is 50% saturated. The P50 value under normal physiological conditions at sea is 26.3 mmHg. The value of P50 may vary according to the physiological conditions or environmental conditions (Fig. 2.3). Shift in oxygen–hemoglobin dissociation curve toward right signifies increased affinity of oxygen to Hb, while a left shift suggests decreased affinity. For example, the P50 is increased in anemic conditions to increase the unloading rate of oxygen from Hb, while it is decreased in case of pulmonary diseases [21, 22].

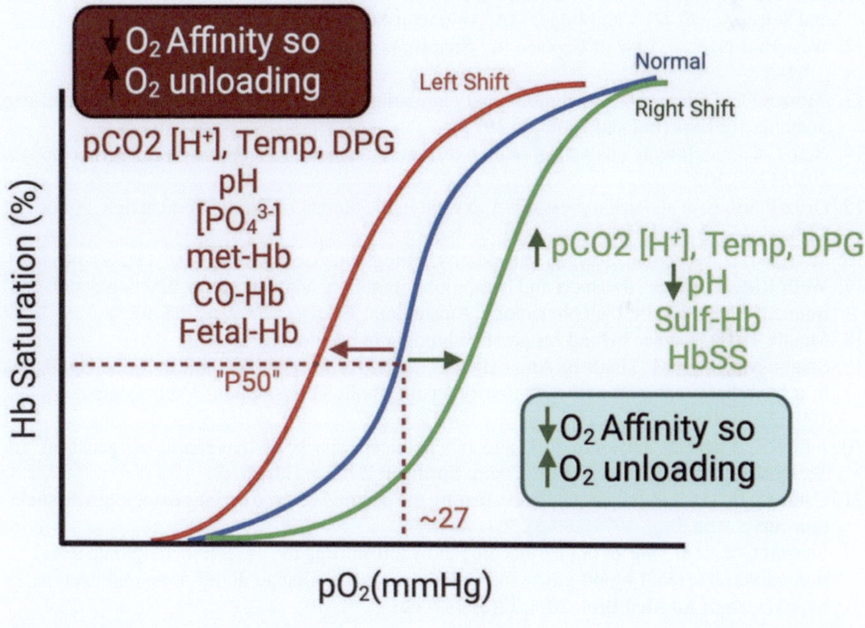

Fig 2.3 Oxygen–hemoglobin dissociation curve. Oxygen–hemoglobin dissociation curve is a sigmoidal shaped curve which represents the cooperative binding of Hb to O_2. As the partial pressure of oxygen increases, the binding of one oxygen molecule increases the affinity of oxyhemoglobin for O_2 in adjacent Hb subunits. The affinity of Hb for oxygen increases or decreases under various conditions resulting in a left or right shift in the oxygen–hemoglobin dissociation curve

References

1. Brimblecombe P. Air composition and chemistry. Cambridge: Cambridge University Press; 1996.
2. Lennard-Jones, J.E. and A. Devonshire, Critical phenomena in gases-I. Proceedings of the Royal Society of London. Series A-mathematical and physical sciences, Royal Society, London 1937. 163(912): p. 53–70.
3. Pavese F, Beciet GMM. Modern gas-based temperature and pressure measurements. Cham: Springer; 2012.
4. Schlatter TW. Atmospheric composition. In: Green aviation. Hoboken, NJ: Wiley; 2016. p. 447.
5. Chandan, G. And M. Cascella, Gas laws and clinical application. StatPearls Publishing, Treasure Island, FL, 2019.
6. Gillespie LJ. The Gibbs-Dalton law of partial pressures. Phys Rev. 1930;36(1):121.
7. Freshwater-Turner D, Cooper R. Physics of gases. Anaesth Intensive Care Med. 2012;13(3):102–5.
8. Laugier A, Garai J. Derivation of the ideal gas law. J Chem Educ. 2007;84(11):1832.
9. He W, Lv W, Dickerson JH. Gas diffusion mechanisms and models. In: Gas transport in solid oxide fuel cells. Cham: Springer; 2014. p. 9–17.
10. Tanaka H. Adiabatic process and the stationary state. Prog Theor Phys. 1955;13(5):497–504.
11. Roebuck J. The joule-Thomson effect in air. In: Proceedings of the American Academy of Arts and Sciences, vol. 60. Cambridge, MA: American Academy of Arts and Sciences; 1925. p. 537.
12. Westerhof N, et al. Law of Laplace. In: Snapshots of hemodynamics. Cham: Springer; 2019. p. 51–5.
13. Altman PL, Dittmer DS. Respiration and circulation. Bethesda, MD: Federation of American Societies for Experimental Biology; 1971.
14. Neto CA, et al. Exergy analysis of human respiration under physical activity. Int J Thermodyn. 2010;13(3):105–9.
15. Ortiz-Prado E, et al. Partial pressure of oxygen in the human body: a general review. American J Blood Res. 2019;9(1):1.
16. Mairbäurl H, Weber RE. Oxygen transport by hemoglobin. Compr Physiol. 2011;2(2):1463–89.
17. Wells RM. Blood-gas transport and hemoglobin function: adaptations for functional and environmental hypoxia. In: Fish physiology. Amsterdam: Elsevier; 2009. p. 255–99.
18. Maulik T. The science behind cooperative binding in haemoglobin.
19. Sharan M, Singh M, Singh B. An analytical model for oxygen transport in tissue capillaries in a hyperbaric environment with first order metabolic consumption. Math Comput Model. 1995;22(9):99–111.
20. Jain KK. Physical, physiological, and biochemical aspects of hyperbaric oxygenation. In: Textbook of hyperbaric medicine. Cham: Springer; 2017. p. 11–22.
21. Canizaro PC, et al. A technique for estimating the position of the oxygen-hemoglobin dissociation curve. Ann Surg. 1974;180(3):364.
22. Stewart GM, et al. Impact of pharmacologically left shifting the oxygen–hemoglobin dissociation curve on arterial blood gases and pulmonary gas exchange during maximal exercise in hypoxia. High Alt Med Biol. 2021;22(3):249–62.

Biochemical and Physiological Aspects of HBOT

3

The discovery of oxygen is attributed to Joseph Priestley in 1772, when he reported the importance of oxygen in combustion as well as in sustenance of life [1]. Oxygen is an elemental gas which exists in atmosphere as a molecule of dioxygen (O_2) and consists of two unpaired electrons making it highly reactive. In biological system, oxygen is involved in the oxidation of energy yielding biomolecules (carbohydrates, lipids, and proteins) and generation of energy in the form of ATP through the process of oxidative phosphorylation [2]. As previously discussed, oxygen is majorly transported in the blood in bound state with hemoglobin and the gaseous exchange of O_2/CO_2 takes place in lungs and tissues due to differential partial pressures of both the gases in these sites.

3.1 Bohr Effect and Haldane Effect

The release of both respiratory gases, i.e. O_2 and CO_2 in the tissues and lungs, respectively, depends on the changes in pH of blood due to their binding with Hb. It can be collectively explained by Bohr–Haldane effect (Fig. 3.1). Bohr effect describes the property of Hb which aids in the release of O_2 in the tissues. Metabolizing tissues release CO_2 in the vicinity, thereby creating an acidic environment. The binding of Hb to O_2 is inversely proportional to acidity and CO_2 concentration facilitating the release of O_2 in the tissues [3].

Haldane effect states that under high acidic conditions Hb binds poorly with CO_2 to form carbhemoglobin. The binding of O_2 to Hb in the blood makes Hb a strong acid. CO_2 is released from Hb to form carbonic acid which readily dissociates into CO_2 and H_2O to be released by the lungs [4]. Combined Bohr–Haldane effect explains the transport and release of O_2 and CO_2 and how pH determines their binding and release [5].

M. Gupta, I. Somasundaram, *Hyperbaric Oxygen Therapy: Principles and Applications*, https://doi.org/10.1007/978-981-99-7278-4_3

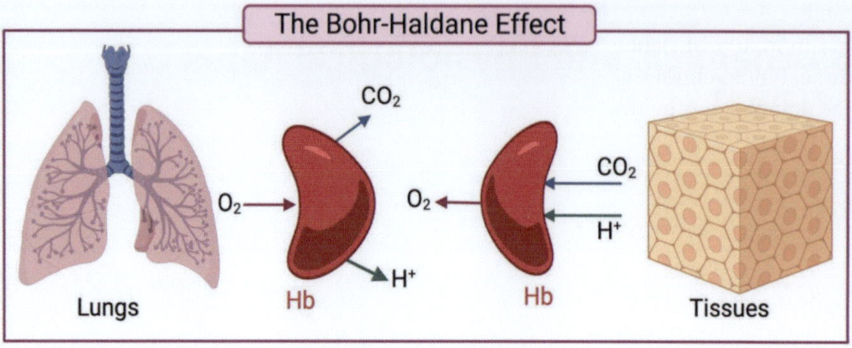

Fig. 3.1 Bohr–Haldane effect. Release of respiratory gases, i.e. CO_2 and O_2 can be explained by Bohr and Haldane effect. Oxygen is released in the tissues in response to decreased pH due to higher CO_2 concentration (Bohr effect), while CO_2 is released in lungs for expiration as the affinity of CO_2 decreases in the presence of higher oxygen concentration resulting in an increase in pH

3.2 Oxygen Consumption

Oxygen delivered to the tissues is rapidly consumed and the rate of oxygen consumption determines the metabolic sufficiency of tissues. Oxygen consumption is calculated as the amount of oxygen utilized by the tissues per minute.

Oxygen consumption (VO_2) is derived by determining the arterial oxygen content (CaO_2), venous oxygen content (CvO_2), and cardiac output, i.e. amount of blood pumped by heart in 1 min (CO) [6, 7].

$$VO_2 = CO \times (CaO_2 - CvO_2)$$

Oxygen consumption rate is dependent on various extrinsic factors such as exercise, trauma, inflammation, and hypothermic conditions.

3.3 Oxygen Debt

Aerobic respiration requires a constant supply of oxygen to fulfill the energy requirements of the body. During periods of intense physical activity like strenuous exercise, there is an increased oxygen demand creating a deficit for oxygen [8]. Under these conditions, anaerobic respiration in the form of lactic acid pathway takes place. There is an accumulation of lactic acid in the muscles which needs to be catabolized into water and carbon dioxide once body comes to a resting state. The additional oxygen or oxygen debt is required for the purpose to restore the state of homeostasis [9]. There are numerous factors upon which oxygen delivery is dependent such as cardiac output, pulmonary gas exchange, oxygen carrying capacity of blood, and rate of oxygen take up by the tissues [10]. During the conditions of oxygen debt, the protective mechanisms in response to cellular stress are triggered. Oxygen debt may be created in surgical and critically ill patients due to hypoxic conditions. This results in hormonal and metabolic changes affecting the morbidity and mortality of patients. The increased oxygen demand needs to be fulfilled by clinical interventions [11]. HBOT has proved to be beneficial for reducing this oxygen deficit and provides oxygen sufficiency to reduce cellular stress.

3.4 Oxidative Phosphorylation and Electron Transport Chain

In the tissues, majority of oxygen is utilized in the mitochondria for the transduction of energy. Mitochondria is a double membraned organelle consisting of energy transduction machinery in the form of multiprotein transmembrane complexes. The high energy substrate in the form of acetyl coenzyme A (acetyl CoA) is derived from the initial breakdown of biomolecules. Acetyl CoA leads to the production of NADH and $FADH_2$ through Krebs cycle [12]. NADH and $FADH_2$ enter the process of oxidative phosphorylation which involves electron and proton (H^+) transport across the inner mitochondrial membrane through five multiprotein complexes named from complex I to V. These five complexes form the electron transport chain (ETC) and the complex I, III, and IV are involved in the translocation of H^+ in the intramembrane space generating an electrochemical gradient and proton motive force (Fig. 3.2). This proton motive force is utilized by complex V for the synthesis

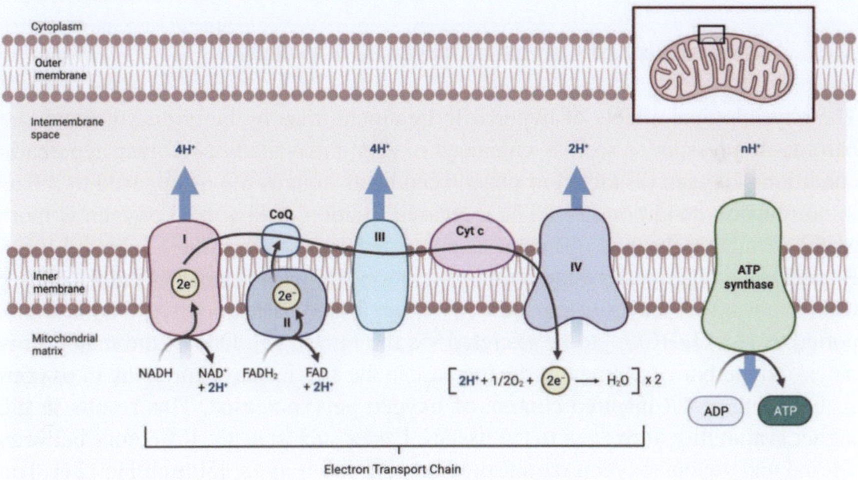

Fig. 3.2 Electron transport chain and oxidative phosphorylation. The energy transduction pathway in cells consists of 4 multiprotein complexes (complex I–IV) and ATP synthase in the inner mitochondrial layer. NADH and $FADH_2$ are formed in the mitochondrial lumen which undergoes oxidative phosphorylation generating electrons and protons. Transport of two electrons takes place via each complex with simultaneous transfer of protons in the inter-membrane space. Oxygen acts as the final electron acceptor. The proton motive force generated by the electron transfer is utilized by ATP synthase to generate ATP

of ATP from ADP and inorganic phosphate (Pi). The final electron acceptor in the electron transport chain is molecular oxygen. Complex IV (also known as Cytc oxidase) transfers 4 electrons and 4 H^+ to oxygen to form a molecule of water (H_2O) [13–15].

3.5 Reactive Oxygen Species and Oxidative Damage

The electrons generated in the electron transport chain are destined to be taken up by oxygen. However, there is always a partial leakage of electrons from ETC. This leads to the production of partially reduced oxygen species known as reactive oxygen species (ROS), namely superoxide radical, hydroxy radical, and hydrogen peroxide [16]. ROS are extremely reactive and have deleterious effects on polyunsaturated fatty acids, nucleic acids, and proteins. One of the most prominent effects of ROS is lipid peroxidation which causes functional and structural damage to the lipid bilayer membranes of cells affecting the functions of membrane associated proteins, channels, and receptors [17, 18]. ROS are produced even in normal physiological conditions but are neutralized by the defense mechanism consisting of cellular enzymes such as superoxide dismutase, glutathione peroxidase, and catalase [19]. Apart from these enzymes, several other non-enzymatic anti-oxidant molecules such as tocopherol, ascorbic acid, and beta-carotene facilitate the elimination of toxic ROS species [20, 21].

3.6 Physiology of Oxygen Utilization in HBOT

The physiological effects of hyperbaric treatment arise by the virtue of increased barometric pressure as well as enhanced oxygen bioavailability. Under hyperbaric conditions, oxygen dissolved in plasma can be as high as 6% as opposed to 3% as in normobaric condition [22]. The increased fraction of dissolved oxygen is more readily available through diffusion as compared to oxygen in the Hb bound form. Hyperbaric oxygen also results in the saturation Hb in the venous blood. However, it does not affect the CO_2 transport in venous blood as majority of CO_2 is transported as H_2CO_3/HCO_3^- to be excreted via the lungs [23]. One of the major attributes of hyperbaric treatment is the increase in the arterial partial pressure of oxygen (paO_2) as the total inspired content of oxygen gets increased. This results in the higher availability of oxygen to the tissues. Under such state the difference between arterial and venous oxygen partial pressure can reach up to 350 mm Hg [24]. The oxygen consumption rate of tissues increases significantly but the rate is dependent on various other factors. The oxygen requirement of tissues varies according to their energy requirement and so the oxygen consumption varies accordingly. For example, the oxygen consumption of cardiac muscles is higher than the skeletal muscles, while the oxygen requirement of skeletal muscles is higher than that of skin tissues [25]. Also, there is a compensatory effect in response to increased oxygen partial pressure in the form of vasoconstriction. Vasoconstriction leads to the reduction in the blood flow to the tissues ensuring optimum tissue oxygenation. There is a decline in cardiac output initiated by vasoconstriction and baroreceptor mediated mechanism [26]. There are various other physiological functions affected by hyperbaric treatment and which are of therapeutic significance. HBOT is known to enhance fibrinolytic activity by decreasing the levels of plasminogen activator inhibitor-I (PAI-I) while increasing the levels of euglobulin fibrinolytic activity (EFA) and tissue type plasminogen activator (t-PA). It is notable that PAI-I level decreases during initial compression, while increase in EFA and t-PA takes place shortly after the decompression [27]. HBOT also has a rejuvenating effect in tissues under stress due to factors including oxidative stress and hypoxia. The higher fraction of dissolved oxygen in plasma aids in increasing the availability of oxygen to the hypoxic tissues or to the cells under stress conditions [28]. Overall, HBOT helps to enhance the availability of oxygen to all the tissues and the fulfillment of oxygen requirement generates secondary responses leading to therapeutic results.

3.7 Nutraceuticals as an Accessory to HBOT

HBOT is recommended in clinical conditions when no signs of improvement are observed in the course of conventional treatment strategies. Administration of HBOT aids in a wide array of functions at physiological as well as cellular level by virtue of increased bioavailability of oxygen to the cells and tissues. HBOT exerts its effect on aspects of wound healing such as reduction of edema and inflammation in addition to augmenting processes of regeneration like angiogenesis, mobilization of stem cells, and extracellular matrix (ECM) deposition [22, 26, 29, 30]. Studies have shown that the beneficial effects of HBOT can be optimized when combined with accessory factors such as utilization of antibiotics and nutraceutical products.

Nutraceutical is a term representing the products derived from food sources with a therapeutic value apart from their nutritional benefits [31]. The term was first coined in 1989 by Stephen DeFelice, founder and chairman of the foundation for innovation in medicine, Cranford New Jersey. On the basis of source of nutraceutical products and their pharmacological composition, they may be classified into four different categories: dietary supplements, medicinal foods, functional foods, and farmaceuticals (farm derived products) [32]. The role of nutraceuticals is mainly considered as preventive medicine aimed to improve the physiological state of the individual. They have the ability to enhance the inherent protective mechanisms of body such as regenerative power, anti-oxidant defense mechanism, immune functions, and gene expression [33–36]. Nutraceuticals aid in the maintenance of physiological integrity, protection against infections, and postponing the process of biological aging.

Nutritional supplements play an important role in effective wound healing during the course of HBOT as well as during the recovery after treatment. Patients suffering from non-healing wounds are adversely affected by the conditions of insufficient nutritional intake [37]. The process of recovery requires regeneration of new tissues which need to be sufficed by providing proper nutritional elements [38].

Nutritional elements which majorly constitute the part of nutraceuticals include polyunsaturated fatty acids, amino acids, keto acids, vitamins, and mineral which plays a key role in improved recovery and well-being of patient undergoing treatment and shown to have enhanced the benefits of HBOT [39].

References

1. Priestley J. Experiments and observations on different kinds of air, vol. 2. London: J. Johnson; 1776.
2. Cramer WA, Knaff DB. Energy transduction in biological membranes: a textbook of bioenergetics. Cham: Springer; 2012.
3. Jensen FB. Red blood cell pH, the Bohr effect, and other oxygenation-linked phenomena in blood O2 and CO2 transport. Acta Physiol Scand. 2004;182(3):215–27.
4. Teboul J-L, Scheeren T. Understanding the Haldane effect. Intensive Care Med. 2017;43(1):91–3.
5. Siggaard-Andersen O, Garby L. The Bohr effect and the Haldane effect, vol. 31. Abingdon: Taylor & Francis; 1973. p. 1–8.
6. Huggett C. Estimation of rate of heat release by means of oxygen consumption measurements. Fire Mater. 1980;4(2):61–5.
7. LaFarge C, Miettinen O. The estimation of oxygen consumption. Cardiovasc Res. 1970;4(1):23–30.
8. Tiep BL, et al. Demand oxygen delivery during exercise. Chest. 1987;91(1):15–20.
9. Brooks GA, et al. Temperature, skeletal muscle mitochondrial functions, and oxygen debt. Am J Physiol. 1971;220(4):1053–9.
10. Ledingham I. Factors influencing oxygen availability. J Clin Pathol. 1977;11:1.
11. Teboul J-L, et al. Value of the venous-arterial PCO2 gradient to reflect the oxygen supply to demand in humans: effects of dobutamine. Crit Care Med. 1998;26(6):1007–10.
12. Nolfi-Donegan D, Braganza A, Shiva S. Mitochondrial electron transport chain: oxidative phosphorylation, oxidant production, and methods of measurement. Redox Biol. 2020;37:101674.
13. Hatefi Y. The mitochondrial electron transport and oxidative phosphorylation system. Annu Rev Biochem. 1985;54(1):1015–69.
14. Chains ET. Oxidative phosphorylation. New York: Academic; 1975.
15. Lodish H, et al. Electron transport and oxidative phosphorylation, in molecular cell biology. 4th ed. New York: WH Freeman; 2000.
16. Nickel A, Kohlhaas M, Maack C. Mitochondrial reactive oxygen species production and elimination. J Mol Cell Cardiol. 2014;73:26–33.
17. Ramana KV, Srivastava S, Singhal SS. Lipid peroxidation products in human health and disease, vol. 2013. London: Hindawi; 2013. p. 1.
18. El-Beltagi HS, Mohamed HI. Reactive oxygen species, lipid peroxidation and antioxidative defense mechanism. Not Bot Horti Agrobot Cluj Napoca. 2013;41(1):44–57.
19. Ighodaro O, Akinloye O. First line defence antioxidants-superoxide dismutase (SOD), catalase (CAT) and glutathione peroxidase (GPX): their fundamental role in the entire antioxidant defence grid. Alexandria J Med. 2018;54(4):287–93.
20. Sies H, Stahl W. Vitamins E and C, beta-carotene, and other carotenoids as antioxidants. Am J Clin Nutr. 1995;62(6):1315S–21S.
21. Sies H, Stahl W, Sundquist AR. Antioxidant functions of vitamins: vitamins E and C, Beta-carotene, and other carotenoids a. Ann N Y Acad Sci. 1992;669(1):7–20.
22. Thackham JA, McElwain DS, Long RJ. The use of hyperbaric oxygen therapy to treat chronic wounds: a review. Wound Repair Regen. 2008;16(3):321–30.
23. Bjerregård A, Jansen E. Monitoring carbon dioxide in mechanically ventilated patients during hyperbaric treatment. Diving Hyperb Med. 2012;1:134.

24. Shinomiya N. Molecular mechanisms of hyperbaric oxygen therapy. In: Hyperbaric oxygenation therapy. Springer; 2020. p. 3–20.
25. Pinnell J, Turner S, Howell S. Cardiac muscle physiology. Continuing Education Anaesth Crit Care Pain. 2007;7(3):85–8.
26. Niinikoski JH. Clinical hyperbaric oxygen therapy, wound perfusion, and transcutaneous oximetry. World J Surg. 2004;28(3):307–11.
27. Weaver LK, et al. Arterial and pulmonary arterial hemodynamics and oxygen delivery/extraction in normal humans exposed to hyperbaric air and oxygen. J Appl Physiol. 2009;107(1):336–45.
28. Lam G, et al. Hyperbaric oxygen therapy: exploring the clinical evidence. Adv Skin Wound Care. 2017;30(4):181–90.
29. De Wolde SD, et al. The effects of hyperbaric oxygenation on oxidative stress, inflammation and angiogenesis. Biomol Ther. 2021;11(8):1210.
30. Schreml S, et al. Oxygen in acute and chronic wound healing. Br J Dermatol. 2010;163(2):257–68.
31. Kalra EK. Nutraceutical-definition and introduction. AAPS PharmSci. 2003;5(3):27–8.
32. Singh J, Sinha S. Classification, regulatory acts and applications of nutraceuticals for health. Int J Pharm Bio Sci. 2012;2(1):177–87.
33. Mishra S, et al. Effects of nutraceuticals on genetic expressions. Open Nutraceuticals J. 2009;2(1):70.
34. Caponio GR, et al. Nutraceuticals: focus on anti-inflammatory, anti-cancer, antioxidant properties in gastrointestinal tract. Antioxidants. 2022;11(7):1274.
35. Cornelli U. Antioxidant use in nutraceuticals. Clin Dermatol. 2009;27(2):175–94.
36. Magrone T, Jirillo E. Disorders of innate immunity in human ageing and effects of nutraceutical administration. In: Endocrine, metabolic & immune disorders-drug targets (formerly current drug targets-immune, endocrine & metabolic disorders), vol. 14. Sharjah: Bentham Science Publishers; 2014. p. 272–82.
37. Tatucu-Babet OA, Ridley EJ. Malnutrition screening in outpatients receiving hyperbaric oxygen therapy: an opportunity for improvement? Diving Hyperb Med. 2018;48(4):206.
38. Alshamali M, et al. Successful post-surgical wound management with hyperbaric oxygen therapy. Asian J Case Rep Surg. 2020;5(2):6–11.
39. Lalieu RC, et al. Nutritional status of patients referred for hyperbaric oxygen treatment; a retrospective and descriptive cross-sectional study. Diving Hyperb Med. 2021;51(4):322–7.

4 Physiological Effects of Elevated Barometric Pressure

Changes in atmospheric pressure result in changes in physiological functions in order to maintain the homeostasis in the body (Fig. 4.1). In natural conditions, an individual is subjected to an elevated ambient pressure when descending below the surface of the sea. It has been determined that a vertical column of 33 ft. or 10 m corresponds to 760 mm Hg pressure equivalent to 1 ATA. Thus, a descend of 10 m in the sea subjects the individual to 2 ATA pressure (corresponding to 1 ATA of air pressure and 1 ATA of pressure exerted by water) [1]. The physiological changes accompanied by the elevation in pressure depend on various factors such as total amount of pressure, duration of elevated pressure, activity level of individual, temperature, rate of descend, and mixture of gases inspired during the course of elevated pressure.

4.1 Effect of Increased Pressure on Physiological Functions

4.1.1 Effects on Hematological Functions

One of the major effects of pressure is development of diuresis and fluid loss resulting in hemoconcentration. The levels of Hb, plasma protein, hematocrit, and cholesterol are found to be slightly increased due to the fluid loss. There is impairment in the sense of thirst in hyperbaric conditions which can worsen the effect of fluid loss. Pressure induced diuresis occurs mainly due to the inhibition of anti-diuretic hormone (ADH) release affecting the hydrostatic action of ADH on renal tubules [2]. Also, the tubular reabsorption of Na^+ ions is inhibited which decreases its active transcapillary movement. There is a steep rise in the concentration of fibrinogen, neutrophil count, and platelet count after exposure to hyperbaric environment. This is usually observed in sea divers where environmental factors such as cold water also contribute to this effect [3]. However, the values gradually return to normal upon returning to normobaric conditions.

M. Gupta, I. Somasundaram, *Hyperbaric Oxygen Therapy: Principles and Applications*, https://doi.org/10.1007/978-981-99-7278-4_4

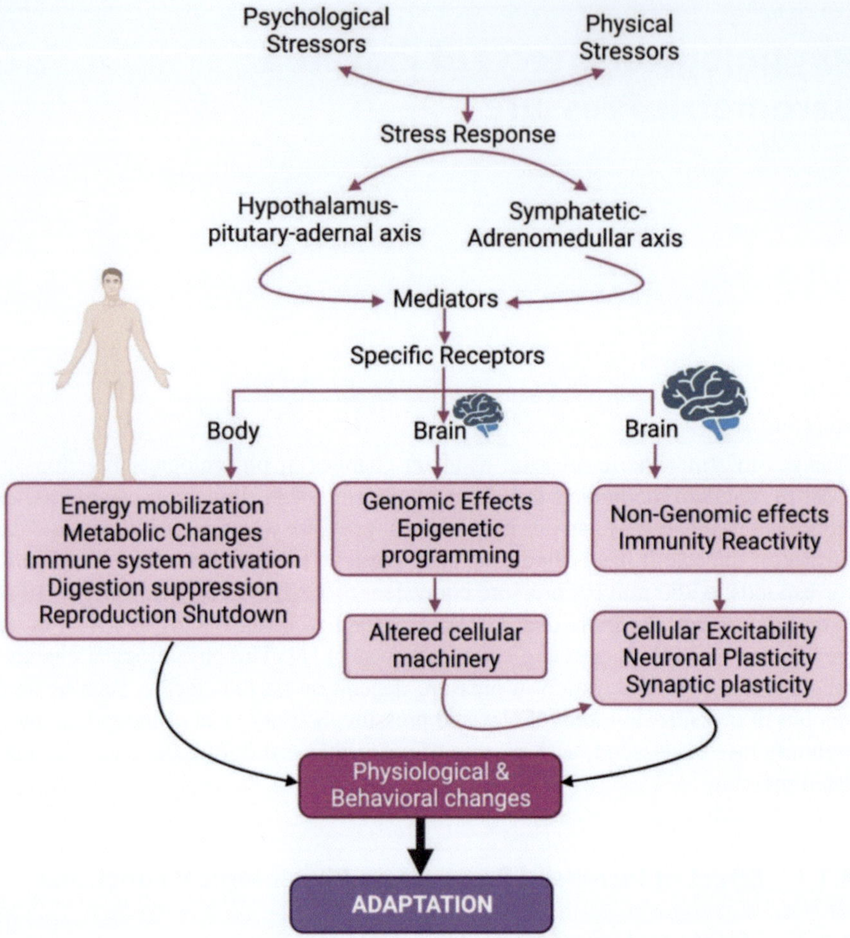

Fig. 4.1 Effect of increased pressure on different physiological functions. Elevation in the barometric pressure affects the hypothalamus–pituitary adrenal axis and sympathetic adrenomedullary axis through stress response. Changes in pressure result in modulation of various physiological functions in order to maintain homeostasis

4.1.2 Changes in Cardiovascular and Skeletal System

Elevation in the atmospheric pressure can interfere with the electrical activity of muscle cells including cardiac muscles. The effects are manifested in the form of mild arrhythmia due to changes in the excitation–contraction coupling. The elevation in pressure results in decreased myocyte excitability and causes conductional instability. However, the extent of conductional instability due to hyperbaric environment has negligible physiological consequences. There is an increase in cardiac contraction with an increase in the left ventricular pressure. The divers exposed to high atmospheric pressure develop circulating gas bubbles resulting in right ventricular overload and impaired ventricular diastolic performance. The elevation in pressure has also been associated with bradycardia. The overall effect of changes in the cardiac functions leads to a decrease in the cardiac output of the individual [4].

There is a generalized vasoconstriction of blood vessels in response to HBOT. Hyperbaric conditions result in an increased production of nitric oxide in the endothelium leading to reduced vasodilation. Function of sympathetic nervous system is also altered in response to elevated pressure promoting vasoconstriction. The partial decrease in blood flow due to vasoconstriction is responsible for the therapeutic effects of HBOT in tissue edema. The cardiovascular implications of HBOT are primarily dependent on the baroreflex mediated mechanism. The state of vasoconstriction triggers baroreflex mechanism via vagal parasympathetic system resulting in bradycardia [5, 6].

Blood flow to the skeletal system is found to be reduced under hyperbaric conditions due to the formation of nitrogen bubbles. The head of femur is mostly affected after prolonged exposure to pressurized air giving rise to dysbaric osteonecrosis characterized by ischemia and subsequent infarction of bone [7].

4.1.3 Effect on Respiratory System

The changes associated with respiratory system upon exposure to hyperbaric conditions are mainly due to altered affinity of Hb to oxygen. This leads to hyperventilation and increased retention of CO_2 in the tissues [8]. The inspiratory and expiratory volumes are affected under pressure where compression leads to hypercapnia, while decompression results in hypocapnia. Due to the increased density of inspired air under pressure there is higher airway resistance and it requires more energy to breathe properly. There is a loss of elasticity of lung tissues upon prolonged exposure to high pressure [9]. However, secondary factors other than hyperoxic conditions are more responsible for the effect. The peripheral oxygen receptors are also affected in their function under prolonged hyperoxic conditions as seen in deep-sea divers [10].

4.1.4 Effect of Other Atmospheric Gases on Physiological Functions

Apart from oxygen, all other atmospheric gases such as nitrogen, helium, and carbon dioxide behave differently under the conditions of high atmospheric pressure. According to Dalton's law, each gas exerts pressure according to the fraction of that gas present in the atmosphere, which is represented by its partial pressure. Breathing compressed air may result in symptoms similar to alcohol intoxication. The narcotic effects arise due to the presence of less soluble nitrogen gas in the air and its differential lipid solubility when inhaled under pressure. The major symptoms of nitrogen induced narcosis include impaired mental ability, decline in cognitive functions, state of euphoria, and impairment in decision making capability. Narcosis resulted due to the increased CO_2 retention in the tissues and hypoventilating conditions. However, the mechanism may be similar to the effect of conventional anaesthetics like nitrous oxide which interfere with the functions of ion channel receptors. The possible disruption of ion channel receptors may lead to neurochemical changes producing the symptoms of narcosis. Helium is less soluble in blood than nitrogen and has a higher diffusion rate. It also has a low fat solubility at normal temperature and pressure. At higher atmospheric pressure, the fat solubility of helium increases affecting mainly the CNS. It has a stimulatory effect on the nervous system and found to be neuroprotective and cardioprotective. However, the mechanism of action of helium and possible effect of elevated pressure on biological system is not clearly understood. Carbon dioxide intoxication may occur under conditions of elevated partial pressure of CO_2 or insufficient expiration of produced CO_2. The patient may experience agitation, headache, and palpitation under CO_2 intoxication. CO_2 narcosis is the condition of CO_2 intoxication at partial pressure of CO_2 more than 6 kPa where patient undergoes tachycardia, experiences air hunger, and may lose consciousness.

4.1.5 Effect on Neurological and Endocrine Functions

Various neuropsychological implications have been associated with the exposure of hyperbaric pressure. At a pressure of 10–13 ATA, decline in motor and neurological functions has been observed. There is an impairment of long-term memory, while short-term memory remains unaffected at high pressure [11]. The neurological impairments may arise due to the complications of decompression sickness but there is no direct evidence for their connection. Middle aged divers are found to develop brain lesions as detected by magnetic resonance imaging (MRI). Most of the effects are temporary and disappear over time [12]. Most divers, when exposed to prolonged period of high pressure develop high pressure neurological syndrome (HPNS). It is associated with a range of neurological symptoms such as tremors, myoclonus, dysmetria, hyper-reflexia, sleep disorders, opsoclonus, and convulsions. Several neuropsychiatric manifestations may also appear in the form of memory impairment, psychoses, and cognitive dysfunctions [13, 14].

Changes in endocrine functions may also be manifested under the influence of hyperbaric conditions. There is an increase in the circulating levels of dopamine, epinephrine, and norepinephrine while a decrease in the levels of ADH without any change in aldosterone secretion. The male reproductive hormones levels are also affected as oligozoospermia has been reported in deep-sea divers. The metabolic functions of individual under hyperbaric condition are affected since the levels of hormones like insulin and thyroxine are altered. The levels of insulin and angiotensin I are increased, while thyroxine levels are decreased in the blood. Apart from this, atrial natriuretic factor (ANF) levels are elevated which is responsible for the diuretic effect of hyperbaric environment [15, 16].

4.2 Pressure Considerations in HBOT

HBOT requires a careful consideration of multiple factors depending upon the treatment requirements, risks involved, and patient's response to treatment. The treatment conditions of HBOT are modulated according to the type of ailment as well as level of severity. The level of pressure and the duration for which the pressure is administered are highly critical in HBOT. The ill-effects of hyperbaric conditions usually develop at a very high pressure as observed in sea divers who may experience pressure as high as more than 10 ATA for a prolonged time period in non-standard conditions [17–19]. In a hyperbaric chamber, the air pressure is increased to a moderate level (2–3 ATA) and treatment is provided in regular intervals instead of exposing the patients to high pressure for long durations. Also, the patients are usually given 100% oxygen which rules out the narcotic effects of other atmospheric gases. The controlled conditions and designing of HBOT modules for treatment ensure the therapeutic utilization of hyperbaric environment while minimizing the side effects of high pressure and hyperoxia.

References

1. Jain KK, Baydin S. Textbook of hyperbaric medicine. Cham: Springer; 2017.
2. Klein L. Alpha-adrenergic (norepinephrine) effect on antidiuretic hormone activity. Investig Urol. 1975;13(2):159–64.
3. Labrouche S, et al. Influence of hyperbaric oxygen on leukocyte functions and haemostasis in normal volunteer divers. Thromb Res. 1999;96(4):309–15.
4. Antigny F, et al. Excitation-contraction coupling and relaxation alteration in right ventricular remodelling caused by pulmonary arterial hypertension. Arch Cardiovasc Dis. 2020;113(1):70–84.
5. Marabotti C, et al. Cardiac changes induced by immersion and breath-hold diving in humans. J Appl Physiol. 2009;106(1):293–7.
6. Marabotti C, et al. Cardiac function and oxygen saturation during maximal breath-holding in air and during whole-body surface immersion. Diving Hyperb Med. 2013;43(3):131–7.
7. Cimsit M, et al. Dysbaric osteonecrosis in experienced dive masters and instructors. Aviat Space Environ Med. 2007;78(12):1150–4.
8. Leach R, Rees P, Wilmshurst P. Hyperbaric oxygen therapy. BMJ. 1998;317(7166):1140–3.
9. Jain KK. The history of hyperbaric medicine. In: Textbook of hyperbaric medicine. Cham: Springer; 2017. p. 3–9.
10. Dean JB, et al. Hyperoxia, reactive oxygen species, and hyperventilation: oxygen sensitivity of brain stem neurons. J Appl Physiol. 2004;96(2):784–91.
11. Ciarlone GE, et al. CNS function and dysfunction during exposure to hyperbaric oxygen in operational and clinical settings. Redox Biol. 2019;27:101159.
12. Knauth M, et al. Cohort study of multiple brain lesions in sport divers: role of a patent foramen ovale. BMJ. 1997;314(7082):701.
13. Talpalar A. High pressure neurological syndrome. Rev Neurol. 2007;45(10):631.
14. Jain K. High-pressure neurological syndrome (HPNS). Acta Neurol Scand. 1994;90(1):45–50.
15. Weihl A. Endocrine responses to elevated ambient pressure. In: Environmental endocrinology. Cham: Springer; 1978. p. 276–82.
16. Popova YA, et al. Effects of various hyperbaric gas mixtures on hormonal parameters of healthy human blood and saliva. Hum Physiol. 2008;34(1):103–13.

17. Edwards ML. Hyperbaric oxygen therapy. Part 1: history and principles. J Vet Emerg Crit Care. 2010;20(3):284–8.
18. Lam G, et al. Hyperbaric oxygen therapy: exploring the clinical evidence. Adv Skin Wound Care. 2017;30(4):181–90.
19. Doran L, Riordan N. Hyperbaric oxygen treatment. J Orthomol Med. 9(4):1.

5 Hyperbaric Chambers: Design and Function

The practice of HBOT requires subjecting the patient's entire body to an environment of elevated pressure and inspiration of 100% oxygen for a specific period of time depending on the treatment. The treatment is provided to the patients in specially designed chambers to maintain the pressure higher than the atmospheric pressure. The conditions inside the hyperbaric chambers can be regulated as per the requirements of treatment. The hyperbaric chambers can be classified according to the capacity to accommodate the number of patients and their basic design (Fig. 5.1). Hyperbaric chambers designed to accommodate a single patient are called monoplace chambers, while the hyperbaric chambers with a capacity to accommodate a larger number of patients are called multiplace chambers. The choice of hyperbaric chamber for treatment depends upon the clinical indications and required hyperbaric treatment protocols. The indications may be chronic and urgent or may be elective which would require different approaches to administer hyperbaric treatment.

5.1 Monoplace Chambers

Monoplace chambers are typically made up of clear acrylic material filled with oxygen where a patient can lie in a semi-recumbent or beach chair position. They are lighter in weight and can be easily installed into existing space. The chamber is filled with 100% oxygen and connected with ports and tubing in order to allow monitoring of physiological parameters and intravenous (IV) supply of fluids or medicine to the patients [1–3]. The patients are usually provided with a mask or hood for intermittent breathing of normal air during the treatment. The mechanism of oxygen supply in a monoplace chamber can be of two types:

1. **Constant purging mechanism:** Oxygen is constantly supplied to the chamber at a specific fixed rate and removed from an outlet to the external environment.

M. Gupta, I. Somasundaram, *Hyperbaric Oxygen Therapy: Principles and Applications*, https://doi.org/10.1007/978-981-99-7278-4_5

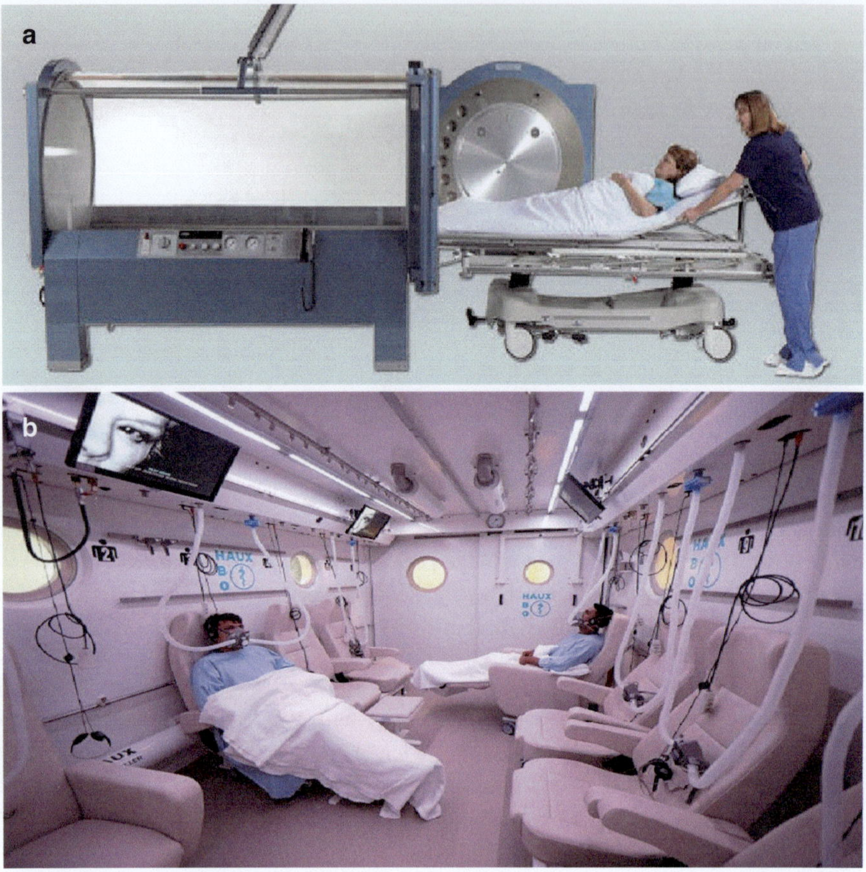

Fig. 5.1 (**a** and **b**) Monoplace and multiplace hyperbaric chambers

2. **Recycling mechanism:** The gases excreted in the chamber environment are recycled by removing unwanted CO_2 and water vapors and supplied back to the chamber.

5.1.1 Advantages of Monoplace Chambers

1. Patients can be treated in an isolated environment minimizing the risk of infection.
2. As the patients stay in the chamber, individual personalized intensive care can be provided without the need to transfer the patient and interrupting the medical treatment.
3. Face mask is not compulsory as the whole chamber environment is saturated with oxygen under pressure. There is less risk of oxygen leakage and the setup is comfortable for the patient.

4. It is easier to observe and monitor the patient during the course of treatment.
5. Decompression procedure is not required.
6. It is economical in terms of space as well as cost and can be easily installed anywhere in the medical facility.
7. The number of operators required to handle the equipment is less.

5.1.2 Drawbacks

1. Higher risk of fire hazard due to completely oxygen filled environment.
2. Medical personnel cannot have direct access to the patient.
3. Physical therapy cannot be conducted due to limited space.
4. For the patients with decompression sickness, it is difficult to provide "air-brake" due to inaccessibility to the patient.

5.2 Multiplace Chambers

Multiplace chambers are specially engineered rooms designed to accommodate a large number of patients (as many as 20 patients). The patients are provided with masks or hoods to receive the supply of 100% oxygen under pressure. Multiplace chambers are spacious and allow medical personnel or caregiver to attend to the patients. Multiple patients can co-occupy the space in a multiplace chamber where specific treatment protocols can be implemented for each individual patient. There are comprehensive monitoring systems to supervise the patient's physiological parameters as well as to check the composition and flow of oxygen given to the patients. The external atmosphere of a multiplace chamber is kept under normobaric conditions with constant maintenance of temperature and humidity [2–4].

5.2.1 Advantages of Multiplace Chambers

1. Multiplace hyperbaric chambers allow simultaneous treatment of large number of patients at a given time.
2. Lower risk of fire hazard.
3. Multiplace chambers can accommodate medical personnel as well as specific medical equipment for the treatment and care of the patients.
4. Physical therapy can be easily provided to the patients.
5. Atmospheric pressures up to 6ATA can be attained in multiplace chambers in case of treatment of specific conditions that require higher pressures such as air embolism and decompression sickness.

5.2.2 Drawbacks

1. One of the major disadvantages of multiplace chambers is lack of patients' isolation subjecting to higher risks of infection and difficulty in providing emergency treatment.
2. Requires high capital, technical expertise, and operating staff.
3. Increased risk of barotrauma during pressurization and decompression.

Hyperbaric chambers can also be classified as soft or hard hyperbaric chambers based on their composition. Soft hyperbaric chambers are made of a soft polymeric material like polypropylene and can be sealed with a zipper. The atmospheric pressure and oxygen levels can only be moderately raised up to 1.3 ATA and 24% O2 level, respectively. These are primarily suited for mountain climbers and divers [5, 6]. Hard hyperbaric chambers are hard-shelled chambers that can achieve higher atmospheric pressures and 100% oxygen levels.

5.3 Safety Measures and Other Considerations for Hyperbaric Facility

The performance and working of hyperbaric facility should be according to the appropriate safety norms for safe and effective administration of hyperbaric oxygen therapy. It is mandatory for hyperbaric centers to follow the norms according to European Code of Good Practice for Hyperbaric Oxygen Therapy. It covers the safety of patients, staff, third parties as well as infrastructure of hyperbaric facility. Following are the major considerations of European Code of Good Practice for HBOT:

5.3.1 Staffing Requirements for Hyperbaric Facility

Hyperbaric facilities should have the following trained staff suited for various aspects of HBOT:

1. Medical director responsible for approval and operation of all the functions in the facility
2. Hyperbaric physician with basic medical qualification and certificate in hyperbaric medicine course
3. Supervisor to oversee all the operations
4. Hyperbaric safety manager to ensure safe practice in the facility
5. Hyperbaric technicians (chamber operator, technicians responsible for repair and maintenance of equipment)
6. Hyperbaric nursing staff (nurse and attendants for direct patient care)
7. Record keeping staff for proper documentation of all the relevant data.

5.3.2 Risk Management and Risk Assessment

Risk assessment involves careful examination of potential hazards in the facility and to ensure that proper precautions and safety measures are in place in order to minimize or avoid injury or harm to patients and staff. This requires proper documentation of procedures, operational conditions (pressure, electrical connections, hygiene, etc.), equipment maintenance logs, and assessment of other factors such as manual errors. The process of risk assessment can be generic but certain specific factors need to be considered for an individual hyperbaric facility.

5.3.3 Procedures

Proper operating manuals for equipment as well as standard operating procedures (SOPs) for various HBOT treatment protocols should always be available to the relevant operating staff. The protocols must have a clear and detailed mention of all operating procedures, emergency protocols, and contingency plans in case of emergency. Apart from this, records pertaining to maintenance of equipment, patient records, records regarding safety, and technical aspects in the facility should always be available to easily track down any discrepancy or technical difficulties.

5.3.4 Patient Safety

Various factors which directly affect patients such as quality of gas, environmental conditions in the chamber, breathing system and apparatus, etc., should have a proper monitoring system and periodical working and maintenance logs of equipment should be kept to avoid any potential hazard to patient safety.

5.3.5 Fire Safety

Proper measures for fire safety should be practiced in the hyperbaric facility. Operational conditions of 1000% oxygen pose a fire hazard, thus additional vigilance and precautionary measures are required in a hyperbaric facility. The electrical connections to the equipment should be of appropriate standard for use in HBOT facility. Use of potential fire hazards like combustible liquids, gases, pressurized cans, spark generating devices, electrostatic generating materials (clothes or any material with electrostatic properties) should be avoided at all costs. Staff should be trained in case of potential fire to ensure their own safety as well as the safety of patients. Patients should be clearly advised about the potential fire hazards. Fire drills and training should be periodically carried out in the hyperbaric center.

References

1. Weaver L. Operational use and patient care in the monoplace hyperbaric chamber. Respir Care Clin N Am. 1999;5(1):51–92.
2. Jain KK. Hyperbaric chambers: equipment, technique, and safety. In: Textbook of hyperbaric medicine. Cham: Springer; 2017. p. 61–78.
3. Lind F. A pro/con review comparing the use of mono-and multiplace hyperbaric chambers for critical care. Diving Hyperb Med. 2015;45(1):56–60.
4. Weaver LK, et al. Arterial and pulmonary arterial hemodynamics and oxygen delivery/extraction in normal humans exposed to hyperbaric air and oxygen. J Appl Physiol. 2009;107(1):336–45.
5. Taber R. Protocols for the use of a portable hyperbaric chamber for the treatment of high altitude disorders. J Wilderness Med. 1990;1(3):181–92.
6. Millar IL. Hyperbaric intensive care technology and equipment. Diving Hyperb Med. 2015;45(1):50–6.

Transcutaneous Oximetry as an Assessment Tool in HBOT

6

6.1 Importance of Transcutaneous Oximetry in HBOT

HBOT administration in various clinical conditions depends on the alleviation of low oxygen tension in the affected region with the manifestations of direct and indirect therapeutic effects. The overall oxygen availability to all the tissues in the body is enhanced during the procedure of HBOT and reflects the health of patient under observation (Fig. 6.1). Thus, it is important to assess the levels of transcutaneous oxygen tension ($TcPO_2$) during the procedure as it reflects the efficiency of HBOT. It is especially beneficial to monitor the transcutaneous oxygen levels in case of treatment of wound healing where local oxygen tension in wounds determines the effectiveness of wound healing under hyperbaric conditions.

Transcutaneous oximetry (TCOM) is a technique to measure the oxygen released from capillaries to the tissues and provides information about the efficacy of oxygen delivery to the tissues. It is a non-invasive and reliable diagnostic method used to determine the oxygenation in the periwound region, i.e. at the edge of wound (Fig. 6.2) [1, 2].

Hypoxic conditions prevail in the wounded tissues and the surrounding region which interferes with the proper healing of wound. HBOT helps in relieving tissue hypoxia and ischemia and improves the conditions of healing. Transcutaneous oximetry is also employed as criteria for selection of patients for HBOT [3, 4]. It is widely used in medical applications other than HBOT such as pediatric intensive care units, plastic surgery, vascular surgery, and orthopedic surgery [5]. Transcutaneous oximetry device uses a Clark type polarographic oxygen electrode placed on skin. The electrode is heated up to 43.5 °C which triggers vasodilation of capillaries underneath skin and gives measurement of oxygen partial pressure of the region (Fig. 6.3) [6].

Under the conditions of hypoxic wounds, the pO_2 becomes less than 40 mmHg. The patients with pO_2 less than 40 mmHg can be considered as potential candidates for HBOT. Under hyperbaric conditions the pO_2 in the periwound region usually

M. Gupta, I. Somasundaram, *Hyperbaric Oxygen Therapy: Principles and Applications*, https://doi.org/10.1007/978-981-99-7278-4_6

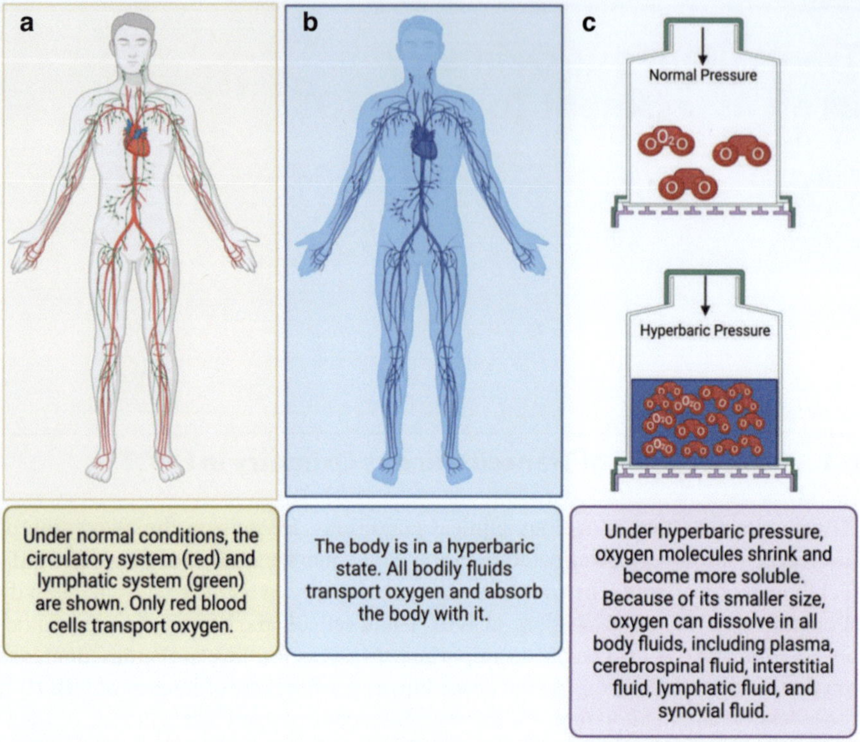

Fig. 6.1 (**a**) Circulatory system (red) and lymphatic system (green) under normal condition. O_2 is transported only by red blood cells. (**b**) Oxygenation in the body under hyperbaric condition. All body fluids carry O2 and saturate the body with oxygen. (**c**) The oxygen molecules become smaller and closely packed under hyperbaric pressure resulting in higher solubility in body fluids

reaches above 200 mmHg [7, 8]. Attainment of higher pO_2 greatly increases the success rate of hyperbaric treatment. There are various factors which may interfere with the measurement of $TcPO_2$ such as edema, acute infection, presence of scar tissue, inflammation, and sclerosis. Thus, a reference $TcPO_2$ reading for the upper thoracic region is also taken along with the measurement of the site of interest [9, 10].

6.2 Algorithms and Considerations in the Measurement of Transcutaneous Oxygen Tension

There are various factors that need to be considered for the application of transcutaneous oximetry in effective management of HBOT procedure. TCOM is a procedure for mapping the oxygen tension of the periwound site. Thus, multiple periwound sites are assessed by TCOM in order to draw a suitable clinical

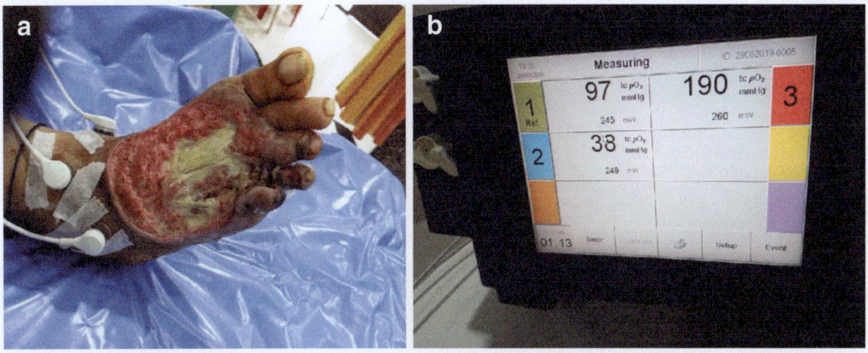

Fig. 6.2 (**a**) A number of leads are placed at locations on the patient's body similar to the setup of an electrocardiogram. (**b**) These leads are connected to transcutaneous oximeter, which measures and records the oxygen measurement from the leads

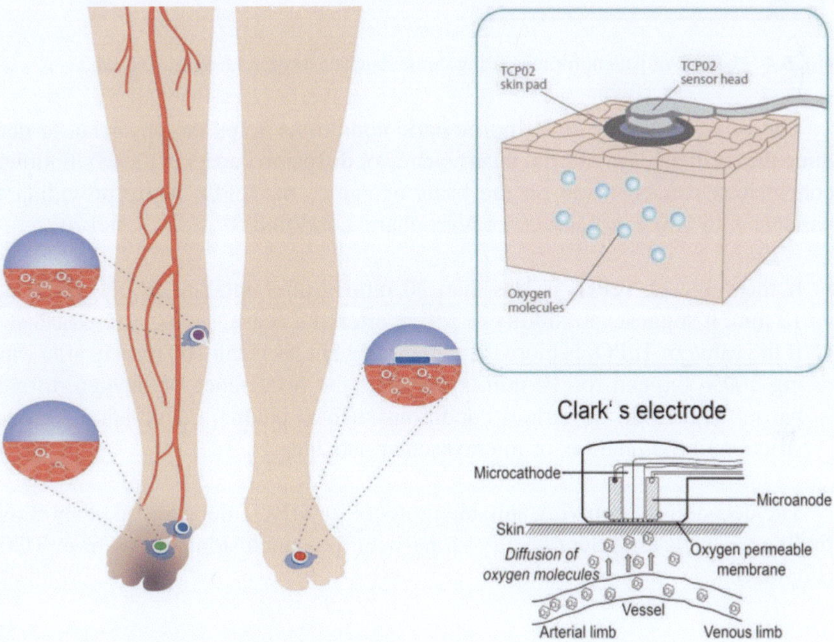

Fig. 6.3 Working of a transcutaneous oximeter

conclusion [11]. The measurements are taken in a warm and comfortable room with patients rested in a supine or recumbent position (Fig. 6.4). The operating temperature of thermistor used in TCOM device is kept at 45 °C. Patients are usually subjected to 100% oxygen under normobaric conditions first to access the variation in their pO_2 before initiating the HBOT. This is called as normobaric oxygen challenge test.

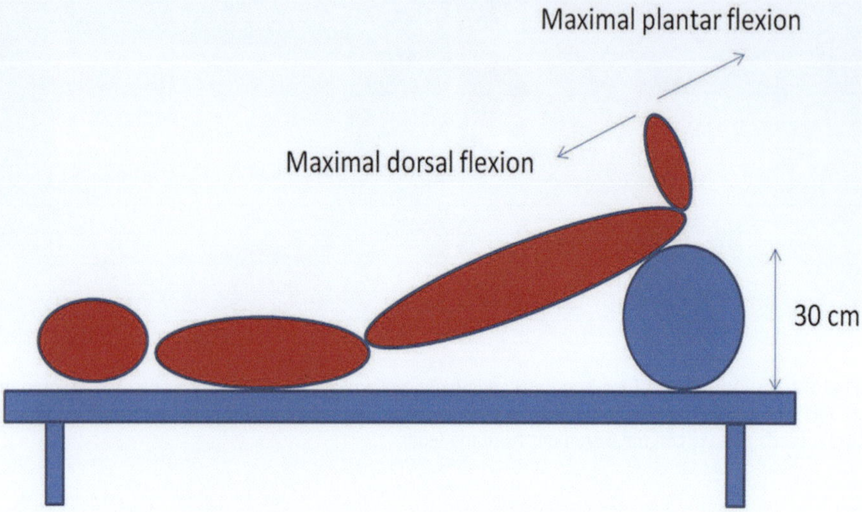

Fig. 6.4 Position of patient for measuring transcutaneous oxygen levels by TCOM

The value of $TcPO_2$ under normobaric conditions helps the physician to determine the incidences of arterial insufficiency or diffusion barriers [12, 13]. Following conclusions can be made on the basis of values of $TcPO_2$ while providing the patients with 100% oxygen under normobaric conditions:

- If the value of $TcPO_2$ is less than 30 mmHg after intaking 100% oxygen for 10 min, it suggests possibility of severe arterial disease.
- If the value of $TcPO_2$ is more than 30 mmHg but less than 100 mmHg after intaking 100% oxygen for 10 min, it suggests the occurrence of oxygen diffusion barrier as a result of various conditions such as edema, dehydration, vasoconstriction, inflammation, or microvascular damage.

The decision for applying suitable protocol for HBOT depends on other clinical findings as well as patient history along with the conclusions drawn from TCOM (Fig. 6.5).

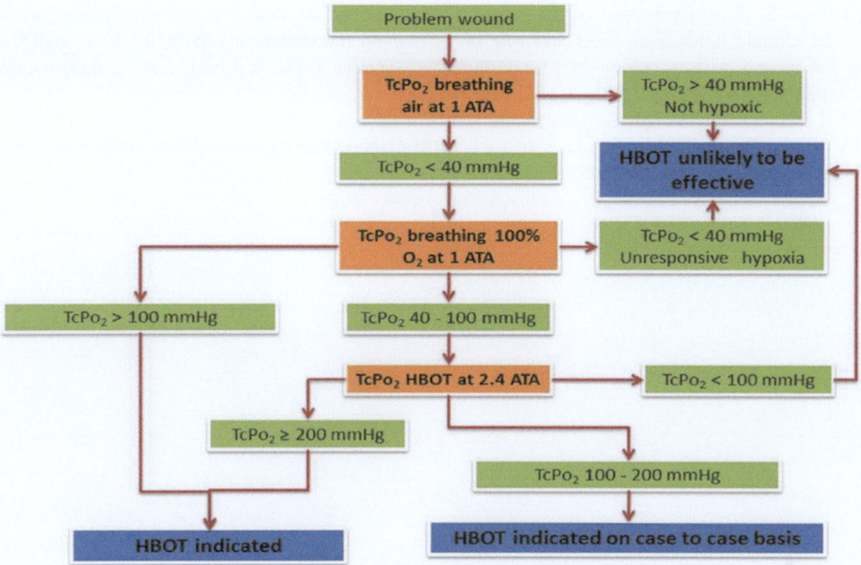

Fig. 6.5 Decision algorithm for transcutaneous oximetry

References

1. Niinikoski JH. Clinical hyperbaric oxygen therapy, wound perfusion, and transcutaneous oximetry. World J Surg. 2004;28(3):307–11.
2. Fife C, et al. Transcutaneous oximetry in clinical practice: consensus statements from an expert panel based on evidence. Undersea Hyperb Med. 2009;36(1):43–53.
3. Arsenault KA, et al. The use of transcutaneous oximetry to predict complications of chronic wound healing: a systematic review and meta-analysis. Wound Repair Regen. 2011;19(6):657–63.
4. Ratliff CR. TCOMs as a screening tool for hyperbaric oxygen therapy. Plast Surg Nurs. 2000;20(1):15.
5. Rooke T. The use of transcutaneous oximetry in the noninvasive vascular laboratory. Inte Angiol. 1992;11(1):36–40.
6. Harsányi G, Péteri I, Deák I. Low cost ceramic sensors for biomedical use: a revolution in transcutaneous blood oxygen monitoring? Sensors Actuators B Chem. 1994;18(1–3):171–4.
7. Smith BM, et al. Transcutaneous oxygen measurements predict healing of leg wounds with hyperbaric therapy. Wound Repair Regen. 1996;4(2):224–9.
8. Niinikoski J. Physiologic effects of hyperbaric oxygen on wound healing processes. In: Handbook on hyperbaric medicine. Cham: Springer; 2006. p. 135–45.
9. Ercengiz A, Mutluoglu M. Hyperbaric transcutaneous oximetry. Treasure Island, FL: StatPearls Publishing; 2017.
10. Leenstra B, et al. Applicability of transcutaneous oxygen tension measurement in the assessment of chronic limb-threatening ischemia. Angiology. 2020;71(3):208–16.

11. Kaur S, et al. Evaluation of the efficacy of hyperbaric oxygen therapy in the management of chronic nonhealing ulcer and role of periwound transcutaneous oximetry as a predictor of wound healing response: a randomized prospective controlled trial. J Anaesthesiol Clin Pharmacol. 2012;28(1):70.
12. Tiffany Hamm B et al. Transcutaneous oximetry.
13. Hauser CJ, et al. Assessment of perfusion in the diabetic foot by regional transcutaneous oximetry. Diabetes. 1984;33(6):527–31.

Evaluating the Recommended Indications of HBOT

7

HBOT has found its place as a therapeutic strategy for a large number of medical conditions. There are irrefutable evidences that support the utilization of HBOT as a standard for treatment in clinical conditions including decompression illness, gas embolism, and carbon monoxide poisoning. Undersea and hyperbaric medical society (UHMS) has approved the use of HBOT for 14 different medical conditions [1]. HBOT is employed as an auxiliary treatment therapy in conjunction with surgery and other clinical interventions for most of the indications and showed an enhanced improvement in these conditions. The range of pathologies include orthopedic, neurological, cardiovascular, gastrointestinal as well as chronic and infective wounds [2–7]. Although HBOT has proved to be beneficial in conditions such as diabetic foot ulcers, ischemic stroke, and multiple sclerosis, further studies with proper control are required to establish HBOT as a standard mode of treatment in these clinical conditions.

7.1 Air or Gas Embolism

7.1.1 Complications of Gas or Air Embolism

Air or gas embolism is the occurrence of air bubbles in the arterial or venous circulation (Fig. 7.1). It is a life-threatening condition which may lead to serious morbidity or mortality. The presence of air bubbles in the vascular system can cause tissue deformation and vessel occlusion affecting tissue perfusion and oxygenation [8]. The presence of gas bubbles in blood also leads to biochemical effects in the form of leukocytes activation, disbalance in homeostasis, and endothelial damage [9, 10]. The contact of gas bubbles with endothelial cells leads to mechanical damage which in turn initiates the opening of transient receptor potential vanilloid (TRPV) ion channels. The opening of TRPV channels allows calcium ion entry, mitochondrial dysfunction, and cell death [11, 12]. Damage to vascular endothelium results in the

M. Gupta, I. Somasundaram, *Hyperbaric Oxygen Therapy: Principles and Applications*, https://doi.org/10.1007/978-981-99-7278-4_7

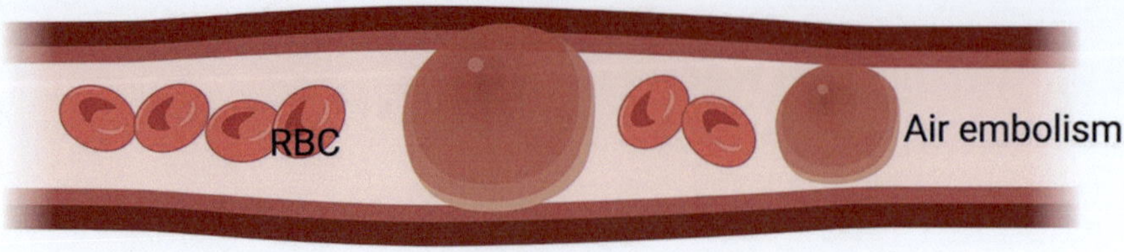

Fig. 7.1 Air embolism. Air embolism is characterized by the presence of gas bubbles in the blood stream leading to the obstruction in blood flow and clumping of blood cells

release of vasoactive substances which severely affects lungs and brain. The injury may be caused due to intracardiac "vapor lock" which results in hypotension or due to circulatory arrest and direct arterial occlusion [13].

Arterial gas embolism (AGE) is well studied in cases of pulmonary barotrauma (PBT) that occurs during pulmonary barotrauma free ascent after compressed gas breathing in submarine escape training [14, 15]. AGE may also occur due to mechanical ventilation, blast injury in or out of water, penetrating chest trauma and medical procedures such as chest tube placement and bronchoscopy [16]. PBT results from gas expansion in the lung cavity during ascent. The clinical manifestations of PBT may occur in the form of local pulmonary injury, pneumomediastinum, pneumothorax, and AGE [8]. Gas bubbles introduced in the pulmonary capillaries enters through left atrium toward ventricle and eventually distributed throughout the vascular system via aorta. The bubbles are passed into the systemic circulation where it causes the occlusion of small and medium sized arteries [17–19].

Venous gas embolism (VGE) is caused by the entry of gas bubbles in the systemic venous system [20]. VGE primarily occurs during invasive diagnostic and surgical procedures. Usually, gas trapped in the pulmonary capillaries do not cause any clinical symptoms but large amount of trapped gas may lead to dyspnea, coughing, and pulmonary edema [21]. VGE may also lead to the occurrence of AGE as the bubbles can enter left atrium and ventricle via atrial septal defect or patent foramen ovale [22]. The pathophysiology of VGE is dependent on the volume of trapped gas in the vascular system and the rate of gas accumulation. Rapid accumulation of gas may cause increase in the right pulmonary artery pressure and right ventricular strain [23]. Pulmonary circulation along with the architecture of alveolar interface enables dissipation of intravascular gases. VGE occurs when there is a rapid introduction of large volumes of gases in the pulmonary blood circulation leading to strain in the right ventricular outflow in addition to increased pulmonary artery pressure and reduced venous return [23, 24]. The clinical manifestations of VGE affects pulmonary, cardiovascular, and neurological system. Cardiac output is reduced during VGE leading to right ventricular failure of insufficiency which in turn result in cerebral hypoperfusion and altered neurological state [25].

7.1.2 Management of air or Gas Embolism by HBOT

The physiological rationale of HBOT for the treatment of gas embolism lies in utilizing the hyperbaric conditions with 100% O_2 or O_2/He mixture to restore the blood flow and provide proper oxygenation to the ischemic tissues [26]. The elevation in the ambient pressure aids in reducing the volume of gas bubbles in the blood circulation. There is an increase in the dissolved plasma oxygen content under hyperoxic conditions during treatment. This helps in adequate oxygen supply to ischemic tissues and buffers the damage to endothelial vasculature due to gas embolism. Under hyperbaric environment, a diffusion gradient is built enabling trapped oxygen and nitrogen to diffuse out of bubbles. Pulmonary and cerebral edema caused due to gas embolism is relieved by vasoconstrictive effect of HBOT [27]. HBOT inhibits the adherence of neutrophils to the post-capillary venule endothelium by antagonizing the β_2-integrin system, thus protecting against reperfusion injury [28].

HBOT is recommended for patients with AGE showing neurological manifestations. Most of the symptoms in patients with AGE resolve after the initial administration of oxygen. However, HBOT is carried out even in the absence of symptoms to resolve the effects of deleterious effects caused due to AGE. The clinical management of AGE using HBOT involves holding the patient in lateral decubitus position or supine position and maintaining a high oxygen concentration using supplemental oxygen. Blood pressure is monitored and maintained during the procedure and hypotensive conditions are treated actively [29, 30]. In case of AGE caused due to PBT, pneumothorax may develop [31, 32]. Thus, it is advisable to treat the patient after chest tube placement in a monoplace chamber. However, multiplace chamber is more preferable due to the ease of monitoring the patient during HBOT. AGE with coexisting pneumomediastinum usually resolves during HBOT and does not require additional therapeutical intervention. HBOT for the treatment of gas embolism is usually carried out at 2.82 ATA at 100% O_2. Earlier treatment strategy as per the U.S. navy Table 6A recommends 6 ATA pressure for 30 minutes but studies have shown that there is no additional effect of using 6 ATA pressure in comparison to using 2.82 ATA pressure. The treatment regimen for gas embolism requires 5–10 HBOT sessions based on the response to treatment. HBOT sessions may be repeated daily or twice in a day if the neurological symptoms persist even after the normal course of therapy [29, 33]. Additionally, various medications can be given in adjunction to HBOT to improve the response to the treatment. For example, Lidocaine is given as an anti-arrhythmic and local anaesthetic agent due to its neuroprotective effects [34]. Similarly, heparin, low molecular weight dextran, corticosteroids and aspirin helps to relieve the secondary symptoms of AGE associated physiological dysfunctions [35].

7.2 Carbon Monoxide (CO) Poisoning

Carbon monoxide is produced in the environment by incomplete combustion of carbon containing materials. CO is endogenously produced in the body by heme oxygenase and functions as a signaling molecule [36]. Exogenous CO exposure to an individual results in symptoms such as unconsciousness, neurological deficits, metabolic acidosis, severe headache, etc. (Fig. 7.2) [37]. Symptoms of CO poisoning are non-specific and variable which often leads to its misdiagnosis as some other ailment. Prolonged exposure to CO affects the normal physiological functions and can prove to be fatal. The primary physiological effect of CO poisoning occurs in the form of impairment in the oxygen carrying capacity of blood leading to tissue hypoxia and its related clinical conditions [38]. However, there are other more direct effects of CO, which are highly toxic to the physiological functioning of body. CO is taken up by lungs and enters the blood circulation where it combines with Hb to form carboxyhemoglobin (COHb). Formation of COHb affects the oxygen carrying capacity of blood and creates hypoxic stress throughout the tissues [39, 40]. CO also binds with hemoproteins such as cytochrome oxidase in mitochondria. Cytochrome oxidase is a part of electron transport chain in mitochondria and its binding with CO affects the energy transduction pathway and promotes the production of reactive oxygen species [41]. CO cause platelets-neutrophil aggregation which leads to vascular and tissue damage [42].

CO has a 200-fold higher affinity for Hb as compared to oxygen [43]. Binding of CO to Hb reduces the arterial content of oxygen and interferes with the release of oxygen to the tissues. The impaired oxygen binding with Hb is reflected in the oxyhemoglobin dissociation curve which shows a left shift indicating a decline in the partial pressure of oxygen in the venous blood (Fig. 7.3) [44]. This suggests that oxygen availability to tissues reduces during CO poisoning. The pathological effects of CO are imparted by its binding to hemoproteins involved in key cellular functions. Competitive binding of CO to key hemoproteins affect their normal physiological functions. For example, nitric oxide free radical (·NO) plays an important role in cell-to-cell communication and other metabolic functions by acting as an intracellular signaling molecule. CO competes with ·NO for the intracellular sites and increases the steady state level of ·NO [45]. The slow dissociation constant of CO affects the overall kinetics of such reactions. Another important aspect of CO poisoning occurs in the form of platelet-neutrophil aggregation in an ·NO dependent mechanism. The production of ·NO increases in response to CO poisoning which in turn generates peroxynitrite ($ONOO^-$) species. Peroxynitrite promotes platelet-neutrophil aggregation by activating platelet adhesion molecules [42].

One of the most adverse effects of CO poisoning is the perturbation of mitochondrial respiration and impairment of energy transduction mechanism. CO can bind with the terminal enzyme of electron transport chain; cytochrome oxidase (CCO) in its reduced state and inhibits the production of ATP disrupting cellular respiration [41]. Additionally, the disruption of cellular respiration is associated with generation of ROS which contributes to cellular stress and apoptosis [46]. CO poisoning also promotes neuronal excitotoxicity by stimulating excitatory neurotransmitters

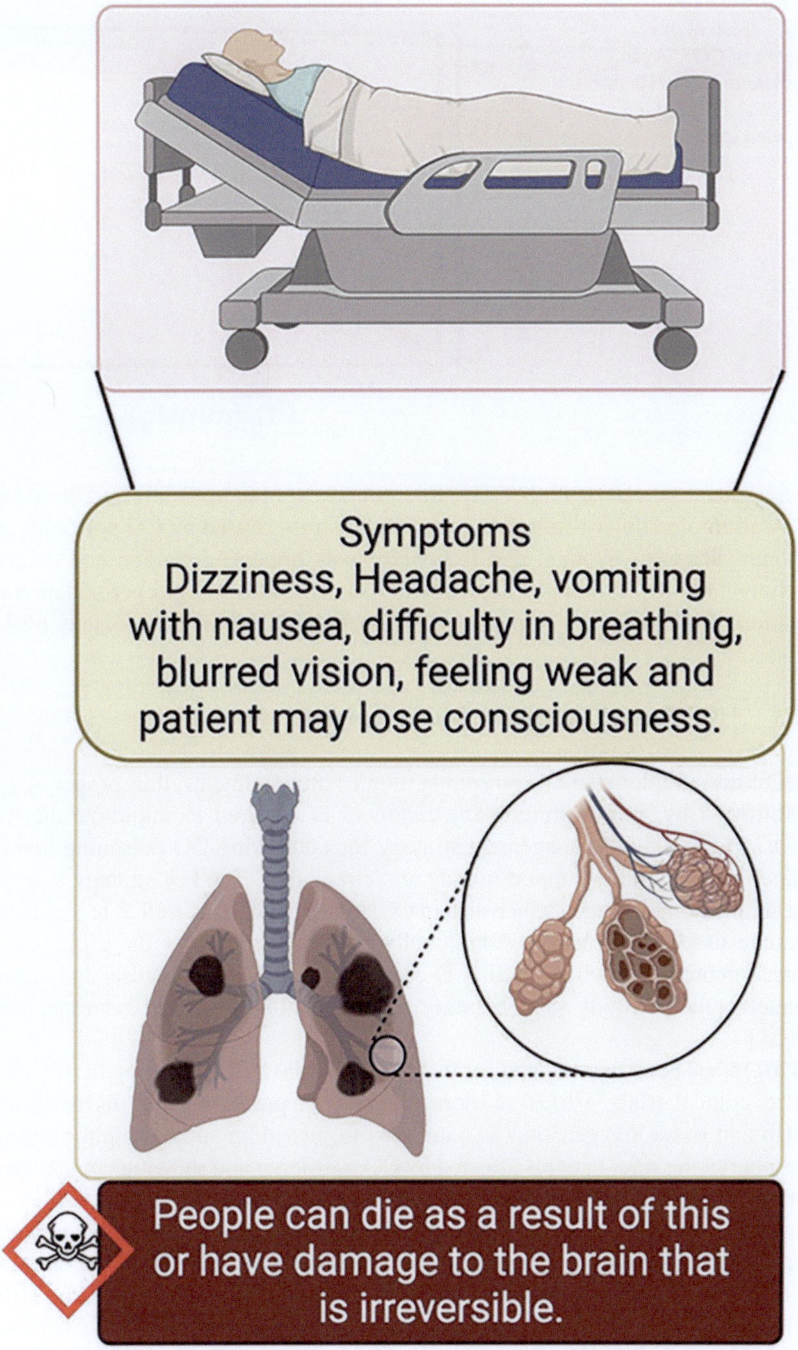

Fig. 7.2 Symptoms of CO poisoning. CO poisoning affects the oxygen carrying function of Hb leading to various symptoms manifesting as dizziness, headache, vomiting, labored breathing, weakness, and loss of consciousness

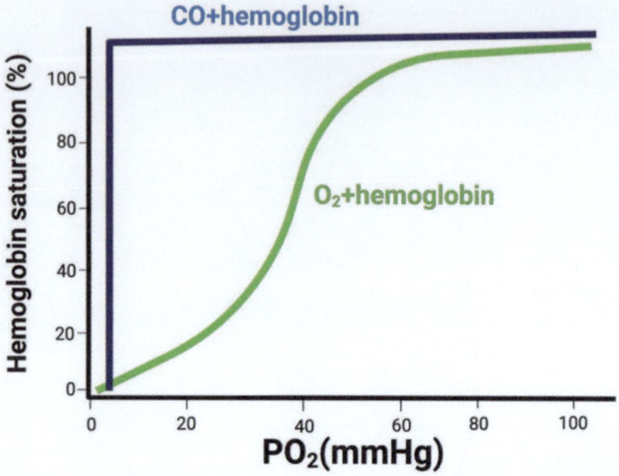

Fig. 7.3 Reduced oxygen affinity of Hb in the presence of CO.CO can readily saturate the Hb due to its greater affinity as compared to O_2

causing neuronal cell death [47]. Heart and brain are the most affected organs due to cessation of cellular respiration and cellular stress caused by CO poisoning. CO poisoning leads to hypoxic and ischemic stress impairing cardiac and neuronal functions [48, 49]. Prolonged exposure to CO also leads to lipid peroxidation and initiation of inflammatory response causing cellular stress and cytotoxicity [49].

7.2.1 HBOT as a Treatment Strategy for CO Poisoning

Most of the symptoms of CO poisoning mimics other ailments, thus proper diagnosis followed by proper preliminary treatment is required to minimize the toxic effects. The standard management strategy for countering CO poisoning involves clinical intervention to support airway and circulation. The key strategy is to provide supplemental oxygen which aids in COHb dissociation as well as re-stablishing adequate tissue oxygenation. Additionally, in order to counter the excitotoxicity induced neuronal deficit, N-methyl D-aspartate (NMDA) antagonists and calcium channels blockers aids in preventing CO poisoning mediated neuronal death [50, 51].

HBOT has been investigated for its efficacy against CO poisoning in many controlled clinical trials. HBOT is more effective in promoting the dissociation of COHb and tissue oxygenation as compared to providing supplemental oxygen at normobaric pressure. Patients affected by CO poisoning and showing clinical symptoms in the form of unconsciousness, neurological problems, cardiovascular dysfunction, and severe acidosis should be immediately treated by providing supplemental oxygen. It is reasonable to access the levels of COHb in the blood and patients with COHb more than 25% should be referred for treatment using HBOT [52]. Usually, a single HBOT session is required to reverse the effects of CO poisoning. However, multiple HBOT sessions can be carried out if required to improve the condition. HBOT treatment protocol for CO poisoning involves an initial compression to 3 ATA which is then reduced to 2 ATA and held for 140 min. This can be followed by two HBOT sessions at 2 ATA for 90 min in an interval of 6–12 h. Conversely, an initial compression of 2.8 ATA followed by administrating 2 ATA pressure for 120 min can be utilized for the treatment. In this case, further HBOT sessions are not required [53–55]. Administration HBOT promotes rapid clearance of CO from blood and tissues maintaining the adequate oxygen levels. The energy transduction system is restored by enhancing the rate of CO dissociation with cytochrome oxidase which helps to preserve the ATP production through cellular metabolism. The oxidative stress caused by the hypoxic conditions is countered by HBOT treatment by upregulation of anti-oxidant enzymes as well as induction of heat shock proteins. Platelet-neutrophil adhesion is inhibited by reduction in myeloperoxidase activity. HBOT provides proper tissue oxygenation which helps to prevent adverse effects of CO poisoning such as lipid peroxidation, inflammation, and tissue necrosis. The prevention of apoptosis and cellular stress helps in the recovery of patients undergoing cardiac and neuronal dysfunctions and ensuring adequate oxygenation to cardiac and neuronal cells [56–58].

7.3 Central Retinal Artery Occlusion (CRAO)

7.3.1 Pathophysiology of Central Retinal Artery Occlusion (CRAO)

Central retinal artery occlusion (CRAO) is a condition of sudden vision loss due to occlusion in the blood supply to the retina (Fig. 7.4). The vision loss is usually caused because of obstruction in the retinal or choroidal blood circulation. The loss of vision is painless and the extent of outcome is dependent on the type of occlusion, i.e., embolus, thrombosis, vasopasm or arteritis. The patients of CRAO often shows retina with a pale appearance due to edema and opacity of nerve fibers and ganglion cell layers [59, 60]. The central retinal artery branches into numerous branches which serves the inner retinal layers. The opthalmic artery gives rise to anterior and posterior ciliary arteries. The long posterior ciliary arteries supply blood to choroid and outer retinal layers while the anterior ciliary arteries serve extra-ocular muscles and anastomose with posterior ciliary vessels to complete the arterial circle of iris. Cilioretinal arteries supplies blood to the retinal region responsible for central vision area, i.e., macula [61, 62]. The presence of cilioretinal arteries in CRAO can save the central vision but the peripheral vision is severely affected. In case of CRAO, the choroidal vessels can supply oxygen to inner retinal layers via diffusion and loss of vision occurs if the retinal tissues are irreparably damaged due to higher degree of occlusion [63]. In most cases of CRAO, natural spontaneous recanalization can occur. The degree of occlusion varies in the patients of CRAO which may result in spontaneous recanalization or recovery by supplying oxygen at normobaric pressure [64]. In other cases, an adequate partial pressure of oxygen is required to initiate recanalization and to keep the retinal tissues viable.

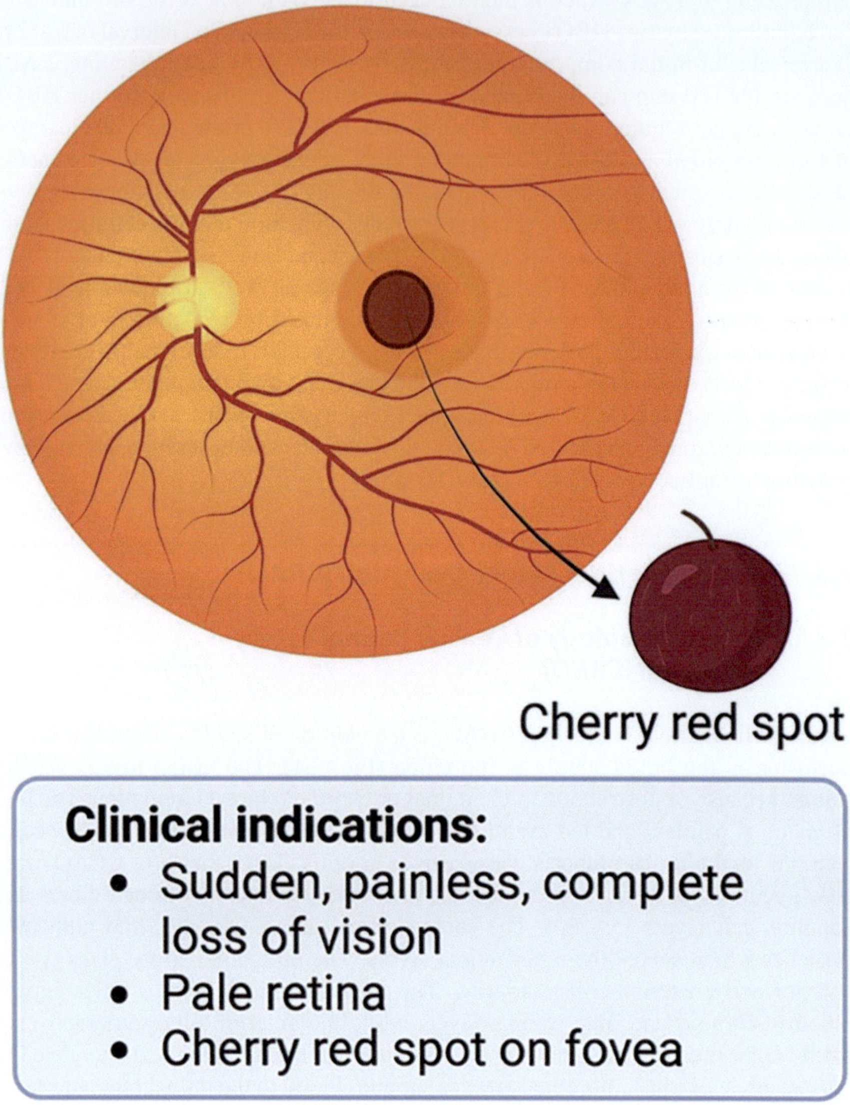

Fig. 7.4 Central Retinal Artery Occlusion. Central Retinal Artery Occlusion is characterized by a painless sudden loss of vision due to blockage of blood flow in one eye. Appearance of cherry red spot can be observed due to retinal edema and thrombus formation

The rate of oxygen consumption of retinal tissues is higher than any other body tissue which makes it highly susceptible to ischemic shock [65]. It is critical to restore oxygen supply to the retinal tissues in order to prevent irreversible damage due to ischemia. Studies carried out on rhesus monkeys after imparting retinal occlusion showed that permanent damage to the retinal tissues occurs if the blood supply is obstructed for more than 105 minutes while recovery is possible if the time duration for occlusion is less than 97 minutes [66]. Treatment strategy for CRAO by using oxygen supplementation either under hyperbaric or normobaric conditions is focussed on ensuring the viability of inner retinal tissues until recanalization and restoration of central retinal artery is achieved.

7.3.2 HBOT in Treatment of CRAO—Strategies and Clinical Evidences for its Effectiveness

The patients should be subjected to the treatment within 24 h of onset of CRAO symptoms because of the sensitive nature of retinal tissues. Treatment beyond this time interval may not be responsive since irreversible damage could occur due to ischemia [67]. The primary treatment strategy involves providing supplemental oxygen to maintain the viability of retinal tissues so that eventually central retinal artery blood supply can be restored. Thrombolytic agents have been used to dissolve embolus or thrombus with limited success [68, 69]. In most of the cases of CRAO, supplemental oxygen can be provided at normobaric pressure to reverse retinal ischemia. There has been a report of a patient with acute vision loss in one eye with findings of CRAO. Initially, the vison was reported to be 20/400 which was improved to 20/25 after administering supplemental oxygen at normal pressure. After discontinuation of supplemental oxygen, the vision immediately returned to 20/400. The patient was kept for 18 hours on supplemental oxygen during which the central retinal artery recanalized and no decline in vision was observed after discontinuation of supplemental oxygen [70]. In another similar case with two patients, significant improvement and restoration of vision was observed after administration of supplemental oxygen at 1 ATA. The vision improved in one patient from 4/200 to 20/70 after 4 h of providing supplemental oxygen while in the second patient vision was restored to 20/200 after having no light perception after 3 h of supplementing oxygen. The condition of improved vision was maintained in both patients after withdrawal of supplemental oxygen [71]. The key factor for the management of CRAO by supplemental oxygen is promptness of treatment. The time factor may vary according to the type of occlusion as well as presence or absence of cilioretinal arteries which aids in keeping retinal tissues viable until recanalization is established. Usually, carbon dioxide is mixed with oxygen to reduce the effect of vasoconstriction due to hyperoxia. However, if choroidal blood circulation is retained, 100% oxygen can be provided since choroidal blood circulation remains unaffected by increased oxygen tension [72, 73]. In another report where both the central retinal artery and temporal posterior ciliary artery was occluded as reported in angiography, the patient was presented after 5 h of visual loss and had minimal light perception. The patient was given carbogen for 10 min every hour around the clock in addition to ocular massage, anterior chamber paracentesis, timolol, and acetazolamide. There was no significant vision improvement after the course of treatment.

However, spontaneous recovery of vision occurred after 96 h from the onset of vision loss [74].

There have been many cases where HBOT is found to be successful in the treatment of CRAO when patients were presented shortly after the onset of symptoms:

(a) Vision improvement in patient treated 90 min after the onset of visual loss was observed after first 10 min of HBOT with subsequent improvement to 20/70 following 5 days of two 90 min HBOT treatments at 2.5 ATA daily for 5 days.
(b) Another patient presented 40 min after visual loss and improved from hand motion to 20/20 after 12 treatments at 2.5 ATA in 9 days.
(c) In another case, the patient presented 4 h after the onset symptoms received 10 HBOT treatments at 2.5 ATA for 90 min each, with gradual improvement from finger-counting vision to visual acuity to 20/30 level.

HBOT is suggested for CRAO when providing supplemental oxygen at normobaric pressure fails to restore the vision. Successful treatment of CRAO by using hyperbaric treatment has been well documented. Treatment of CRAO by HBOT for 16 patients demonstrated significant improvement in 11 patients. In the remaining patients with no improvement, four patients were delayed for more than 24 h in starting the therapy. The patients were treated at 2.0 ATA pressure with 100% oxygen for a duration of 90 minutes initially for 2–3 days. The treatment frequency was reduced to one session per day until clinical plateau is reached [75]. In another study for HBOT intervention in CRAO, 17 patients were divided into four groups based on the time at which their treatment was initiated from the onset of symptoms of CRAO. It was found that HBOT was most effective when the treatment is administered within 8 h from the onset of symptoms of CRAO [76].In a matched control study of treatment with HBOT, 35 patients treated with HBOT were compared for outcome with 37 patients treated under normobaric conditions. The group of patients with HBOT were given oxygen at 2.8 ATA for 90 min twice a day initially for 3 days and reduced to 1 session per day after further improvement was not observed. There was improvement in vision of 82% patients in the HBOT group as compared to 29.7% in normobaric group [77]. The current reports and findings regarding the treatment of CRAO with HBOT suggests that utilizing hyperbaric conditions instead of normobaric conditions has additional effects in reducing the apoptotic state in ischemic retina and effectively maintaining its viable state.

7.4 Idiopathic Sudden Sensorineural Hearing Loss (ISSHL)

7.4.1 Idiopathic Sudden Sensorineural Hearing Loss (ISSHL)

Idiopathic sudden sensorineural hearing loss (ISSHL) or sudden deafness implies a sudden loss of hearing ability or rapidly progressive hypoacusis which is mostly unilateral and, in some cases, an immediate cophosis. The hearing impairment results in a loss of at least 30 dB which occurs usually within 3 days. The clinical

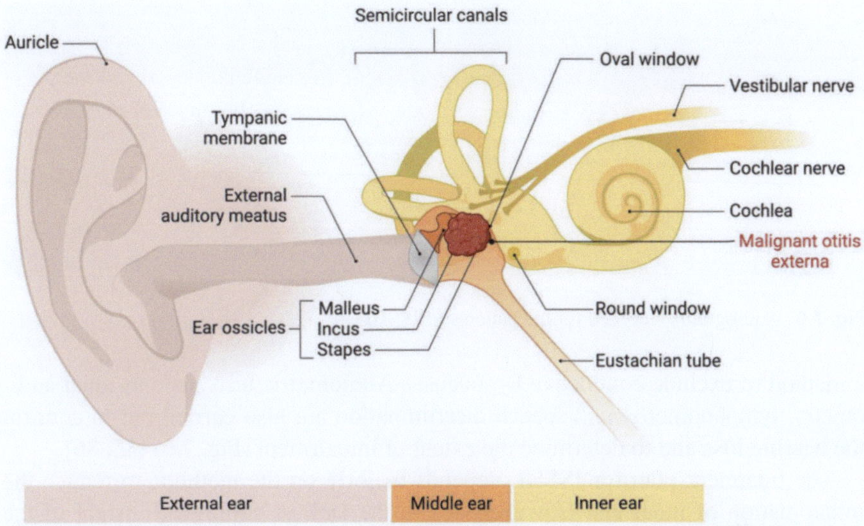

Fig. 7.5 Idiopathic sudden sensorineural hearing loss. ISSNHL is a unilateral sudden loss of hearing over 30 dB at least three contiguous frequencies and usually accompanied by tinnitus and vertigo

manifestations of ISSHL includes unilateral hearing loss, aural fullness, tinnitus, and vertigo (Fig. 7.5). The condition of hearing loss is termed as "idiopathic" since the cause of hearing loss upon clinical and laboratory investigation cannot be identified [78, 79].

The hearing loss in ISSHL occurs due to damage to cochlea. The clear cause of cochlear damage and onset of ISSHL could be determined in only 20% of cases [80]. The etiology of ISSHL can be attributed to one of the following factors:

- Vascular—Thrombosis or embolus of arteria labyrinthi can lead to hearing loss due to reduced oxygen partial pressure in the inner ear [81].
- Viral—Viral infections such as mumps, rubella, and varicella can cause ISSHL as a result of viremia leading to edema in the intima of inner ear [82].
- Round window rupture—Physical trauma to the inner ear or increase in the intercranial pressure which can cause rupture of ear round window may lead to ISSHL [83].
- Auto-immune disorders—Auto-immune diseases such as lupus erythematosus, Cogan's syndrome, Buerger's disease, etc. may lead to cochlear inflammation which in turn can cause ISSHL [84].

The hearing loss in ISSHL is mostly unilateral and its manifestation may be related to pre-existing medical conditions such as diabetes, hypertension, dyslipidemia, and arteriosclerosis [80]. The diagnostic examination for ISSHL includes neurological vestibular and otoscopic examinations in addition to tuning fork test as

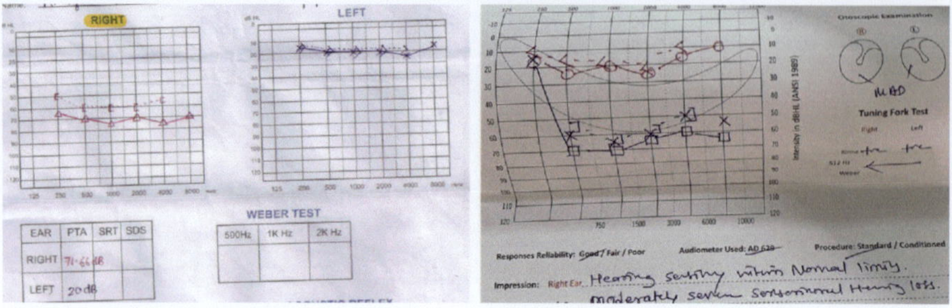

Fig. 7.6 Audiograms recorded for the patients of ISSHL

a method to exclude conductive hypoacusis. Audiometric tests such as tonal audiometry, tympanometry, and speech discrimination are also carried out to confirm the hearing loss and to determine the extent of impairment (Fig. 7.6) [85, 86].

The treatment plan for ISSHL depends majorly on the methods to reduce the inflammation of inner ear. However, due to the lack of etiological insight of the condition it is troublesome to design specific therapeutic strategy against ISSHL. Wide range of therapeutic agents have been suggested for the treatment of ISSHL such as vasodilators, hemodilution agents, steroids, metabolic activators, anti-viral drugs, vitamins, and HBOT. Corticosteroids are immediately given to the patients suffering from ISSHL beginning with 1 mg/kg/day over a period of 2–3 weeks. It is difficult to establish the effectiveness of therapeutic interventions as the outcome is mostly similar to the spontaneous recovery rate in most cases [87–89].

Cochlear activity is directly affected by the rate of oxygen consumption as the stria vascularis and the organ of corti have a high metabolic activity requiring high oxygen demand. It is observed that the perilymphatic oxygen tension decreases in patients with ISSHL. As a result, the sensory neuroepithilium suffers damage due to anoxia or resulting edema which increases the labyrinthine fluid pressure [90, 91]. Administration of supplemental oxygen has been reported to alleviate the symptoms of ISSHL restoring the cochlear function. The metabolic demand of intracochlear region can be met by increased output of pentose phosphate pathway during oxygen administration. HBOT helps in increasing the perilymphatic oxygen pressure resulting in increased amount of available dissolved oxygen and vasoconstriction of cochlear arteries which leads to reduced edema [91]. With improved oxygenation of inner ear, the transmembrane potential and ATP synthesis is restored as well as the electrophysiological function in the labyrinth is maintained by activation of Na^+/K^+ pump restoring the ionic balance. Improved electrical activity in the cochlea is observed after the administration of HBOT. The metabolic functions of cochlea are restored after HBOT by promoting the aerobic oxidation in the stria vascularis and glycolytic anaerobic respiration in the organ of corti. Oxygen inhalation under hyperbaric conditions aids in increasing oxygen tension in perilymphatic fluid and decreasing the hematocrit and blood viscosity. HBOT helps against hypoxic conditions in the cochlear region and restores the functional state of cochlea [92–94].

7.4.2 Evaluation of HBOT as a Treatment Strategy for ISSHL

HBOT has been tested for the treatment of ISSHL and showed significant improvement in the condition by increasing intra-cochlear oxygen tension and oxidative metabolism. The recommended protocol for HBOT against ISSHL is providing patients with 100% oxygen at 2–2.5 ATA for 90 minutes per day. The overall treatment regime can be carried out for 10–20 sessions [95, 96]. Meazza et al. reported the efficacy of two cycles of 15 HBOT sessions after 8 days from the onset of ISSHL [97]. Aslan et al. evaluated the effectiveness of HBOT in 50 patients divided into two groups of 25 each. The first group was treated with betahistidine hydrochloride, 1 mg/kg/day of prednisone and daily stellate ganglion block with lidocaine. The second group received HBOT at 2.4 ATA for 90 minutes for 20 sessions in addition to the same treatment. The mean hearing ability was improved significantly for the second group (37.90 ± 24.0 dB in HBOT group as compared to 20.0 ± 19.6 dB in first group) but the patients older than 60 years showed no additional outcomes [98]. Similar results were reported by Racic et al. in a study on 115 patients. First group of 51 patients received HBOT alone at 2.8 ATA for 60 minutes twice daily while the second group of 64 patients were given intravenous pentoxyfylline. The mean hearing ability in HBOT group (46.35 ± 18.58 dB) was found to be higher as compared to pentoxifylline group (21.48 ± 13.15 dB). Persistent outcomes were reported in both the groups after following up 9 months later [99]. In another study by Fattori et al., two groups of patients treated in an early stage of ISSHL (within 48 hours) showed significant statistical difference in HBOT group as compared to group of patients that were given vasodilator in terms of mean global hearing gain [100].

7.5 Wound Healing

7.5.1 Physiology of Wound Healing

Wound consists of loss of integrity of intact tissue structures as a result of damage from mechanical, physical, thermal or chemical source. Wound healing is an interplay of various processes which work in coordination to repair the damaged tissues. Local environment factors and systemic factors also influence the process of wound healing [101, 102]. Non-healing wounds are a major concern to public health as traditional surgical procedures may not suffice to solve the problem and most of the patients suffering from non-healing wounds are not surgical candidates. In case of non-healing wounds such as diabetic ulcers, the process of wound closure is often countered by a high rate of wound dehiscence [103, 104]. Various non-surgical modes of treatment are under development to deal with the problem of non-healing wounds.

The process of wound healing involves interaction of multiple cell types and biochemical functions. Upon injury to the tissues, platelets and fibrin are localized to the site of wound. This is followed by the movement of fibroblasts, microphages, endothelial cells, and smooth muscle cells to the wound site. Cytokines are released to promote the cellular growth and production of extracellular matrix. Dead and necrotizing tissues are engulfed and cleared off by phagocytic cells. This is followed by the production of collagen and angiogenesis. The wound site is replenished with new cells until the wound is completely healed [101, 102].

Oxygen plays an important role in the process of wound healing. It is involved in various signaling pathways and acts as a cofactor in enzymatic processes. Additionally, leucocyte mediated killing of bacterial cells and collagen formation requires oxygen. The metabolic rate of cells in the wound healing site increases significantly as higher rate of cellular respiration and oxidative clearance of contaminants increases the oxygen demand of these tissues [105, 106]. Hypoxic conditions in case of lower extremities ulcer leads to impaired wound healing process and providing supplemental oxygen at higher pressures aids in augmenting the process of wound healing [107].

Wound healing process has four distinguished phases as hemostasis, inflammation, proliferation, and remodeling (Fig. 7.7). All of the above phases are intermingled and interdependent [108, 109]. Non-healing wound results from the interruption in any of the phases of wound healing due to impairment in the tissue repair. Initially, ROS mediated vasoconstrictions of vessels at the site of injury helps in reducing the blood flow to the wound and initiates platelet and thrombin activation create blood clot. This results in reduction in local perfusion and platelet degranulation releasing various chemical signals such as platelet derived growth factor (PDGF), transforming growth factor-β (TGF-β), epidermal growth factor (EGF), and insulin like growth factor-1 (IGF-1). Mast cells migrate to the wound site and releases bradykinin, complement factors, and histamine resulting in perturbation of microcirculation. ROS release also mediates the mobilization of neutrophils to the wound for protection against bacterial infection. ROS stimulates endothelial cell division, keratinocyte division, fibroblast division as well as migration to the site of injury to promote angiogenesis and ECM formation. Cytotoxic cells such as neutrophils and macrophages act against microbial infections by utilizing ROS mediated phagocytosis. Increased uptake of oxygen by these cells leads to a respiratory burst where ROS production is upregulated through NADPH dependent pathway [101, 110, 111]. Thus, in order to maintain microenvironment for effective wound healing adequate supply of oxygen is critical. Lower partial pressure of oxygen results in a reduced rate of mitochondrial energy production creating hypoxic conditions which negatively affects the process of wound healing. Macrophages and lymphocytes also release lactate and various growth factors like IGF-1, interleukin 1, interleukin 2, TGF-β, and vascular endothelial growth factor (VEGF) to promote endothelial cell division and inflammation. Neutrophils and macrophages releases proteases such as elastase, collagenase, metalloproteinase, and metalloelastase to degrade the

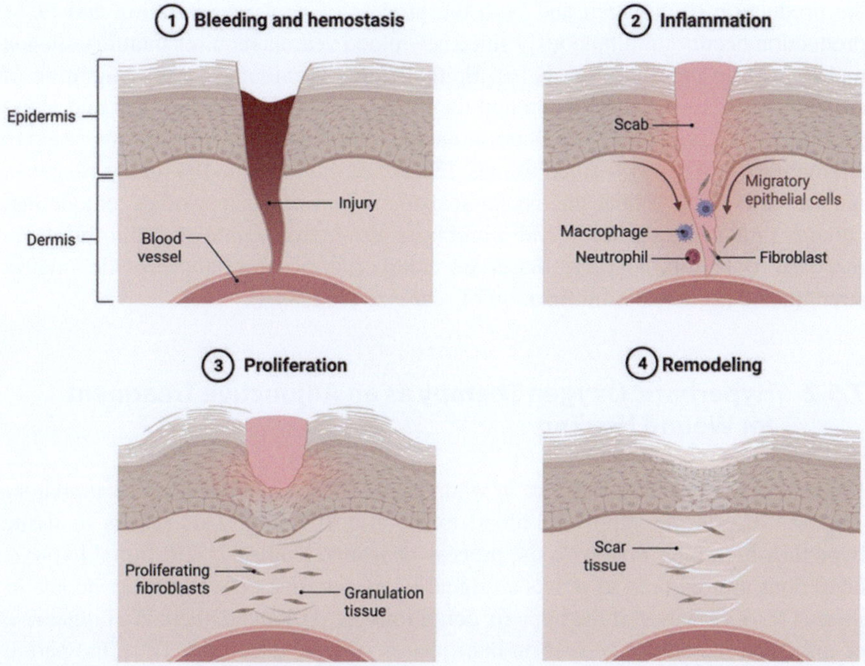

Fig. 7.7 Phases of wound healing. The process of wound healing consists of orchestration of complex phases with characteristic cellular and molecular events. The phases of wound healing include bleeding and hemostasis, inflammation, proliferation, and remodeling. Hemostasis involves cessation of bleeding and formation of thrombus and fibrin mesh after the injury. Inflammation phase prepares the wound bed and mainly focuses on the removal of cellular debris and bacteria from the site of wound. Proliferation phase involves filling of wound bed with granulation tissue and migration of epithelial cells and proliferative fibroblasts for covering the wound. In the maturation phase strengthening and reorganization of collagen fibers help in increasing the tensile strength of scar tissues to complete the healing process

damaged ECM [112]. The microenvironment of the wound site becomes acidic due to an increased production of lactate [113]. The oxygen is mainly utilized in the production of ROS and lactate which generates hypoxic conditions. Hypoxia acts as a signal to repair the damaged tissues but also leads to higher risk of infection and creates poor healing conditions [114].

The proliferative phase begins after 4 days of injury and consists of granulation, new tissue formation, and epithelization. Angiogenesis (generation of new vessels from pre-existing vessels) and vasculogenesis (formation of new blood vessels from endothelial precursor cells) occurs to replace the damaged vasculature at the wound site. Neovascularization occurs in response to oxidative stress, hypoxia, and high lactate concentrations [108]. Tissue hypoxia triggers the release of VEGF which in turn stimulates the release of endothelial precursor cells (EPC) form the bone marrow. EPC undergoes differentiation under the influence of various growth factors to create new blood vessels [115]. Lactate, hypoxia, and growth factors also stimulate the production of collagen and both the process of neovascularization and ECM production occurs simultaneously since new blood vessels requires mature collagen matrix to form their network [116]. Epithelization is initiated by the migration of epithelial cells to the wound site and their replication in the wound bed and along the edges of the wound [117]. Maturation and remodeling of healing wound help in strengthening of newly formed tissues. Fibroblast migration to the matrix assists in matrix remodeling to bear the mechanical stress. During this period of remodeling, collage synthesis continues and fibroblasts differentiate into myofibroblasts in response to TGF-β. Myofibroblasts are contractile in nature and provides matrix strength and promotes minimization of scar size [118, 119].

7.5.2 Hyperbaric Oxygen Therapy as an Adjunctive Treatment for Wound Healing

Damage to the vascular structure in wounds and edema as a result of inflammatory response affects the supply of blood to the site of injury. This results in tissue hypoxia which interferes with the process of wound healing [120]. Initial hypoxic conditions are required as it acts as signal to the immune cells to act on the site of injury [114]. However, if the hypoxic conditions are prolonged, there is a higher risk of infection and adverse effect on later phases of wound healing. Thus, the partial pressure of oxygen in the healing tissues needs to be adequate for proper wound healing.

Oxygen delivery to the tissues depends on the volume of blood supply and depends on diffusion. Thus, if the arterial partial pressure of oxygen is high, availability of oxygen to the tissues also increases. In this scenario, the levels of Hb does not act as a limiting factor. HBOT is beneficial in this condition as it elevates the arterial and capillary oxygen tension as the overall amount of oxygen in the blood is raised during HBOT [6]. Under normobaric conditions the partial pressure of oxygen in the blood is limited to the content of oxygen in the atmospheric gases as per Henry's law. HBOT enables the increase in the partial pressure of blood oxygen to a comparatively higher levels as the total dissolvable oxygen levels are increased (Fig. 7.8). HBOT is advised as an adjunct therapy to accelerate the process of wound healing and it is especially beneficial in the treatment of problem wounds or non-healing wounds. The recommended conditions for HBOT administration include subjecting the patients to 2–2.5 ATA pressure with 100% for 90 minutes continued for 10–40 sessions [121–123].

HBOT affects the process of wound healing at multiple levels as it improves tissue oxygenation (Fig. 7.9). Hyperoxic conditions leads to vasoconstriction which results in decreased in flow of blood to the site of injury. As a result of this, tissue edema is remarkably reduced [121]. Higher availability of oxygen to the tissues improves cellular metabolism and ATP production for the critical functions of the healing tissues. Immune cells like neutrophils and macrophages require higher amount of oxygen to generate ROS to exert their cytotoxic effects against microbial infection.

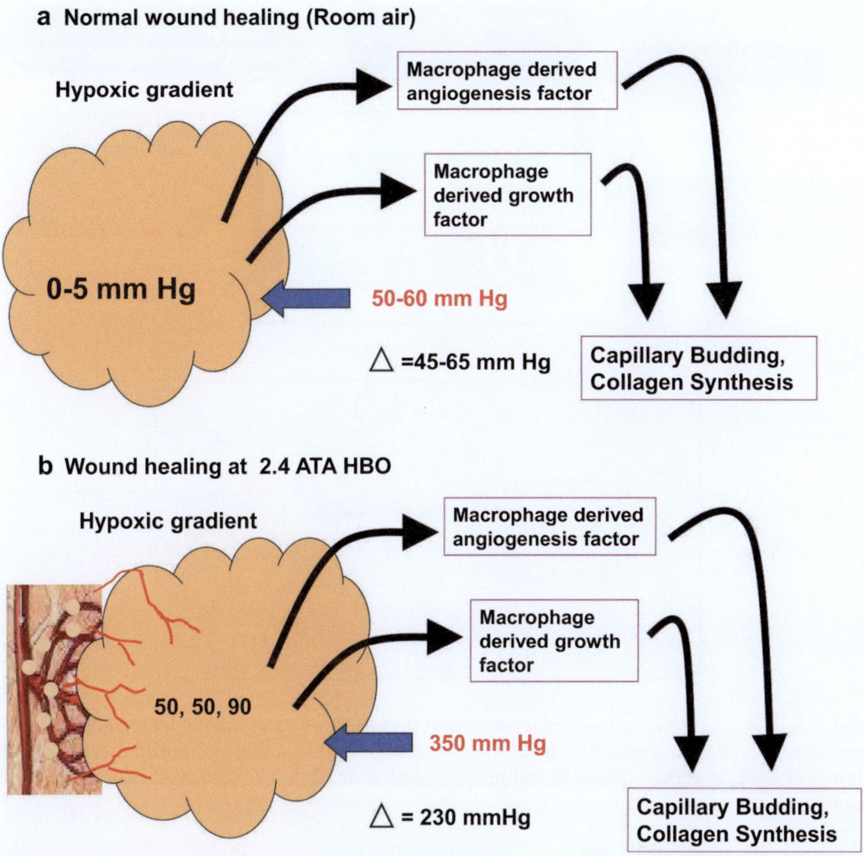

Fig. 7.8 Comparison of wound healing in (**a**) Normobaric condition and (**b**) Hyperbaric condition. Normal wound healing requires a minimum oxygen gradient of 50–60 mm Hg for healing to take place while the oxygen gradient during hyperbaric oxygenation can be as high as 350 mm Hg

HBOT provides adequate oxygen which aids their ability to perform bactericidal and bacteriostatic functions [124, 125]. HBOT also helps neovascularization and collagen formation as both these processes are oxygen dependent and under excess oxygen availability there is an enhancement in these functions. Collagen synthesis involves hydroxylation which is an oxygen dependent process [126]. It is also found that HBOT improves the effectiveness of certain antibiotics such as cefazolin [127]. HBOT enhances the secretion of growth factors such as bFGF, PDGF, and TGF-β1 [128]. EPCs are mobilized form the bone marrow to the site of injury to generate new blood vessels. HBOT stimulates their mobilization via eNOS dependent pathway [115].

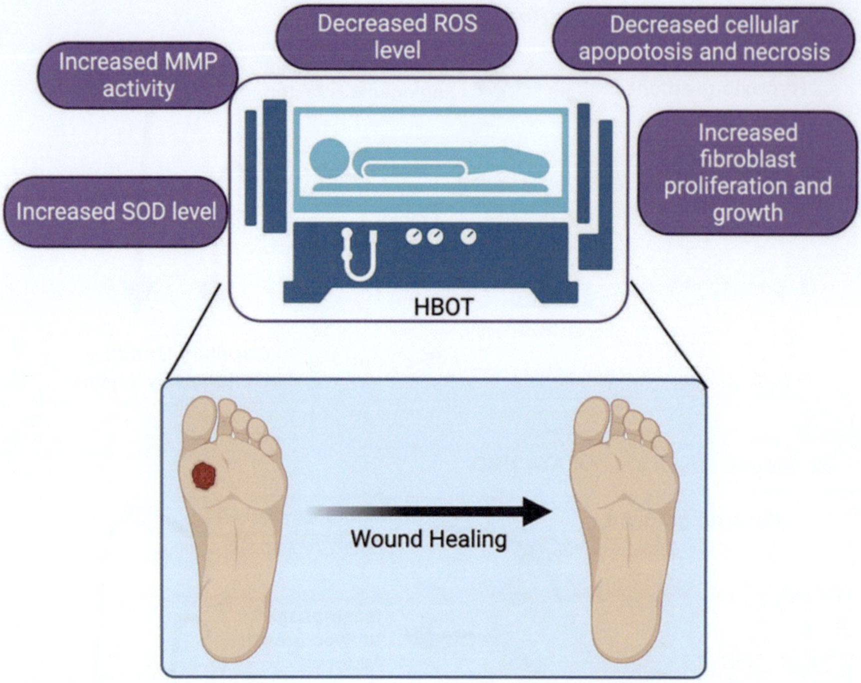

Fig. 7.9 Effect of HBOT on wound healing. HBOT enhances wound healing by modulating various functions such as increasing the SOD level, increasing the activity of MMP, decreasing the levels of ROS, lowering apoptosis and necrosis, and increasing the proliferation and growth of fibroblasts

7.6 Arterial Insufficiency—Diabetic Foot Ulcers and Problem Wound

7.6.1 Problem Wounds or Non-healing Wounds

The occurrence of long-term non-healing wounds poses a great challenge to the healthcare system as they constitute "problem wounds" which tend to persist and fails to heal within a reasonable time frame [129]. Treatment of problem wounds requires specific nursing care, frequent hospitalization visits, and medical attention. Diabetic foot ulcers, radiation, cold, and thermal burns are some of the examples of wounds which do not heal easily. As discussed earlier, wound healing is a complex multi-step process and each phase of wound healing requires interplay of many factors [101]. In case of non-healing wounds, one or more phases of wound healing are impaired. Various processes such as ability to fight infections, regeneration of new cells, and tissues or deposition of extracellular matrix may get affected which hampers the normal wound healing process.

Wound healing requires an array of complex processes such as establishment of homeostasis, inflammatory response, mobilization of immune cells and mesenchymal cells, angiogenesis and collagen formation [101, 109]. There may be multiple factors involved in the occurrence of non-healing wounds which could be intrinsic or extrinsic. Two of the most common factors that affect the normal healing of wound are infection and ischemia. If the wound is infected, there is an elevated inflammatory response associated with tissue edema. Inflammatory conditions and edema lead to hypoxia which in turn interferes with wound healing. The inflammatory phase gets prolonged to combat infection which delays the healing process. Ischemia occurs due to damage to blood vessels and tissue edema. This also results in hypoxic conditions decreasing the availability of oxygen to the healing tissues. The oxygen requirement of tissues at the wound site is higher than the normal tissues as they are actively metabolizing in order to regenerate and heal [130–132]. Additionally, phagocytic cells require excess amount of oxygen to fight infection by generating higher quantities of ROS. There are other processes like fibroblast proliferation which is dependent on the presence of oxygen. Collagen synthesis also requires oxygen for the hydroxylation of lysine and proline to provide tensile strength to the synthesized collagen. Angiogenesis, on the other hand, relies on hypoxic conditions but the maturation of newly formed blood vessels also requires normal levels of oxygen [105, 106]. Direct assessment of oxygen availability to the periwound skin and an indirect oxygen measurement of periwound microcirculation can be carried out by measuring transcutaneous oxygen tension ($TcPO_2$) [133].

Large numbers of primary pathological condition can lead to non-healing wound or problem wound. This includes venous or arterial insufficiency, diabetes, and pressure induced wounds. The common etiological factors in occurrence of non-healing wounds in all these conditions comprises of tissue hypoperfusion, ischemia, hypoxia, and infection [134]. Ulcers of lower extremities and resulting amputations from diabetic ulcers is a serious problem in diabetic patients and the resulting wounds of non-healing nature need to be managed for proper healing [134]. Control of infection in the wound, surgical debridement, enabling vascular sufficiency and extremity offloading are some of the measures to manage diabetic foot ulcers [135]. Adjunctive therapeutical interventions are required for an effective diabetic foot ulcer management.

Ischemia, infection, and sensory neuropathy are the major factors responsible for the pathology of diabetic foot ulcer. Ischemia results in weakening of local defence mechanisms against infection due to the lack of oxygen and nutrient supply through blood. $TcPO_2$ value of 30 mm Hg is considered as a critical threshold for predicting the severity of diabetic foot ulcer. Ischemic and neuropathic conditions often promote infection of wound [136]. The severity of infection can be graded from a mild localized infection to necrotizing fasciitis. Wagner Grading System is widely used for the classification of diabetic foot ulcers on the basis of extent of infection (Table 7.1) [137]. Apart from tissue injury, bone and joint healing are important factor for the proper healing of wound.

Table 7.1 Wagner grading system for diabetic foot ulcers

Grade 0	Intact skin
Grade 1	Superficial without penetration deeper layers
Grade 2	Deeper reaching tendon, bone or joint capsule
Grade 3	Deeper with abscess, osteomyelitis or tendonitis extending to those structures
Grade 4	Gangrene of some portion of the toe, toes /forefoot
Grade 5	Gangrene involving whole foot or enough of the foot that no local procedures are possible

7.6.2 HBOT for the Treatment of Non-healing Wounds

Since hypoxia is one of the major governing factors in the etiology of non-healing wound, oxygen supplementation under enhanced pressure conditions proves to be beneficial in the treatment of non-healing wound (Fig. 7.10). TcPO2 value is often considered as a criterion to assess the periwound hypoxia. HBOT helps in elevating the arterial partial pressure of oxygen (PaO_2). The rate of oxygen consumption by the tissues, capillary blood flow, inter-capillary distance, and arterial oxygen pressure determines the capillary oxygen pressure. The overall tissue oxygenation is enhanced promoting the conditions for wound healing [138, 139]. The inter-capillary distance is increased in wound site due to edema and the ischemic conditions are prevailed due to vascular damage. HBOT has been used in promoting the healing of different types of wounds and holds importance especially in non-healing wounds. In the initial stages of healing hypoxia helps in initiating the process of wound healing. However, later stages are predominantly governed by the higher oxygen requirement for immunological, inflammatory, and regenerative functions [105]. HBOT affects the systemic delivery of oxygen and enhances its diffusion gradient in subcutaneous tissues up to 10–20 times the normal value promoting hyperoxygenation. The major hyperoxic effect of HBOT comes from the increased amount of plasma dissolved oxygen which gets readily diffused to the hypoxic sites. There is a marked improvement in the function of leucocytes, reduction in edema, enhanced angiogenesis, upregulated fibroblast activity, formation of granulation tissues, and accelerated collagen synthesis under the influence of HBOT [140–143]. Additionally, for the regeneration of new tissues, there is an increased mobilization of stem cells from the bone marrow under hyperoxic conditions [144]. HBOT facilitates the delivery of oxygen to the hypoxic tissues during early stages of healing which helps in reducing the adherence of neutrophils and excess release of ROS [145]. This helps in the decrease of tissue necrosis and keeps the tissues viable during the course of healing.

Available evidences regarding the use of HBOT in the treatment of non-healing wounds suggests its potential as an adjunctive treatment strategy. This is according to the requirements for American hyperbaric association (AHA) class 1 based on level A evidence of positive randomized trials. The treatment protocols for HBOT vary as per the type of hyperbaric chamber to be used as well as severity of the condition. HBOT for non-healing wounds is recommended at 2–2.5 ATA of 100%

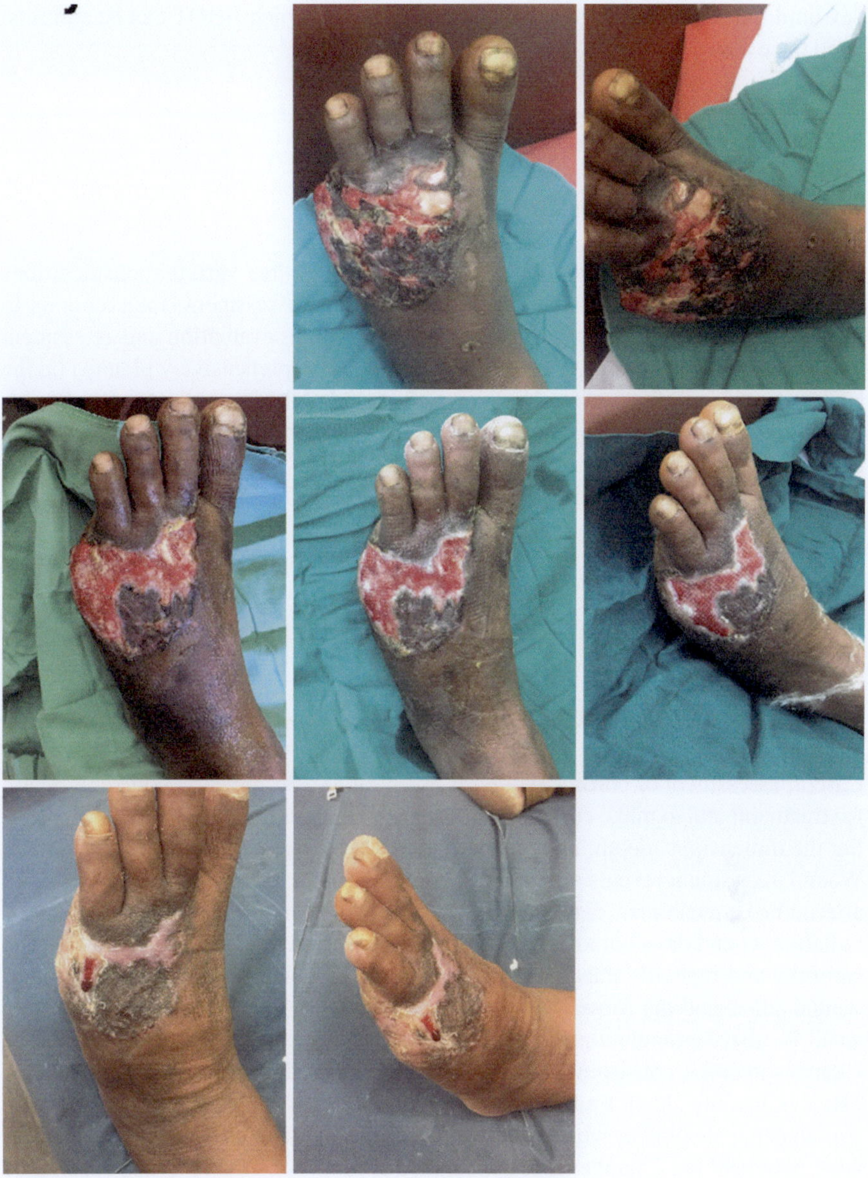

Fig. 7.10 Effect of HBOT on diabetic foot from day 1 to day 30

oxygen for 90–120 min [146, 147]. Other medical interventions in the form of intravenous antibiotics, surgical procedures and measures to control diabetic conditions may be required along with HBOT [148]. Treatment of problem wounds requires monitoring and management of other comorbidities which usually affects the normal process of wound healing. HBOT sessions are advised to be carried out twice a day until the condition of the patient stabilizes; after which HBOT can be administered once a day.

7.7 Chronic Wounds

7.7.1 Management of Chronic Wounds

Management of chronic wounds is complex and requires various considerations with respect to patient care. It demands meticulous assessment, interventions to manage physiological and social factors, and continuous evaluation and assessment of outcomes. The type of interventions needs to be comprehensively planned on the basis of careful evaluation of physiology of wound and contributing pathophysiology. Chronic wound management is based upon the understanding of psychological, social, and behavioral aspects of the patient as well as possible ramifications of chronic wound. Various complications may arise regarding pain management, psychological discomfort, and financial limitations, which must be taken into account [149]. There are other factors such as functional assessment, nutritional status, immunocompetence, and possible requirement of assistive devices which need to be monitored along with the regular laboratory and clinical assessment.

7.7.2 Assessment of Chronic Wound and Wound Environment

Careful assessment of chronic wound is essential to determine the course of action for treatment and to make correct diagnosis. Detailed examination includes measuring the dimensions, depth, and anatomical features of wound and periwound area. Wound assessment reveals the type of chronic wound in the form of pressure ulcer, arterial or venous ulcer, diabetic neuropathic foot ulcer, soft tissue necrosis, post-radiation wound or osteoradionecrosis [150]. The progress of healing should also be carefully and regularly checked. Healing wound is indicated by the contraction of wound edges and the formation of granular tissues. In the early stages of healing, small localized granulation area can be observed which are addressed as islands. Changes in color, consistency, adherence to wound bed also gives an indication of extent of healing [151]. Epithelization of wound edge which appears as pearly with a mounded, pale pink or white is indicative of healing wound while its absence suggests poor healing. Characteristic appearance of periwound tissue is also considered as indications of wound healing. There are various clinical studies and imaging techniques that are employed to assess the nature of chronic wound such as color Doppler, pulse volume recording, ankle brachial indices, tissue biopsy, wound culture, transcutaneous oximetry, and X-rays [152, 153].

Both the microscopic and macroscopic environment of wound gives an account of the progress of wound healing. General appearance of the wound in terms of its texture, consistency, color, and odor constitutes the macroscopic features while the presence of epithelial and immune cells, released proteases and cytokines, microbial load in the wound are primarily considered as microscopic features [154]. The general criteria for proper healing condition are to keep the wound clean, moist, and protected. Local or topical agents may be used to maintain the clean and moist condition of wound. The micro- and macroenvironment of the wound should not be negatively affected by these agents. Similarly wound dressing should maintain the moisture, support the healing microenvironment, and protect the wound and peri-wound tissues. Infected state of wound should be managed by using different treatment options in the form of topical antibiotics, sharp debridement, systemic antibiotics and using wound dressings with biologically active silver. There may be other factors which may alter the healing state of wound such as age of patient, comorbidities, other injuries, and nutritional status [155].

7.7.3 HBOT for the Management of Chronic Wounds

Management of chronic wound often requires a combinative approach comprising of systemic treatment to improve local wound environment and therapeutic strategies to treat underlying pathological conditions. HBOT is often considered as an adjunctive treatment strategy which aids greatly in reinforcing healing conditions for wounds. HBOT has been used as a treatment modality against chronic wounds for nearly 40 years and has shown significant effects in healing of wounds [156]. Typical HBOT involves pressure administration of 2–2.5 ATA for a period of 60–120 min once or twice in a day. The course of treatment is carried out for 15–30 days depending upon the improvement in the condition of the patient [157].

The process of wound healing is known to be dependent on the presence of oxygen. Oxygen plays an important role in a myriad of functions pertaining to wound healing [105]. In the microenvironment of the wound, hypoxic conditions prevail due to vascular damage and edema. In case of acute wound, initial hypoxic condition is beneficial which promote low pH and high lactate concentration to facilitate anti-microbial functions and triggering pathways to mobilize specific cells to the wound area [114]. However, in chronic wound conditions, hypoxia leads to poor regional perfusion to wound hampering the healing process (Fig. 7.11). HBOT administration enables high arterial oxygen partial pressure allowing adequate perfusion to the wound. Hyperoxygenated state is achieved by HBOT which promotes various functions involved in wound healing such as vasoconstriction, fibroblast activation, anti-microbial action, anti-inflammatory effect, release of growth factors, collagen formation and deposition, reduced leucocyte chemotaxis, and enhance the action of antibiotics [57, 119, 126, 127, 141, 142].

Due to the toxicity of oxygen to the tissues, hyperoxygenated conditions can only be sustained for short period of time and the treatment strategy involves intermittent hyperoxic-hypoxic conditions. The repeated cycles of hyperoxic-hypoxic condition create hyperoxia-hypoxia paradox and enable favorable environment for the processes of wound healing dependent either on higher or normal oxygen demand (Fig. 7.12). Clinical evidences suggest that such intermittent exposure is beneficial in wound healing. Adequate oxygen tension persists even after the

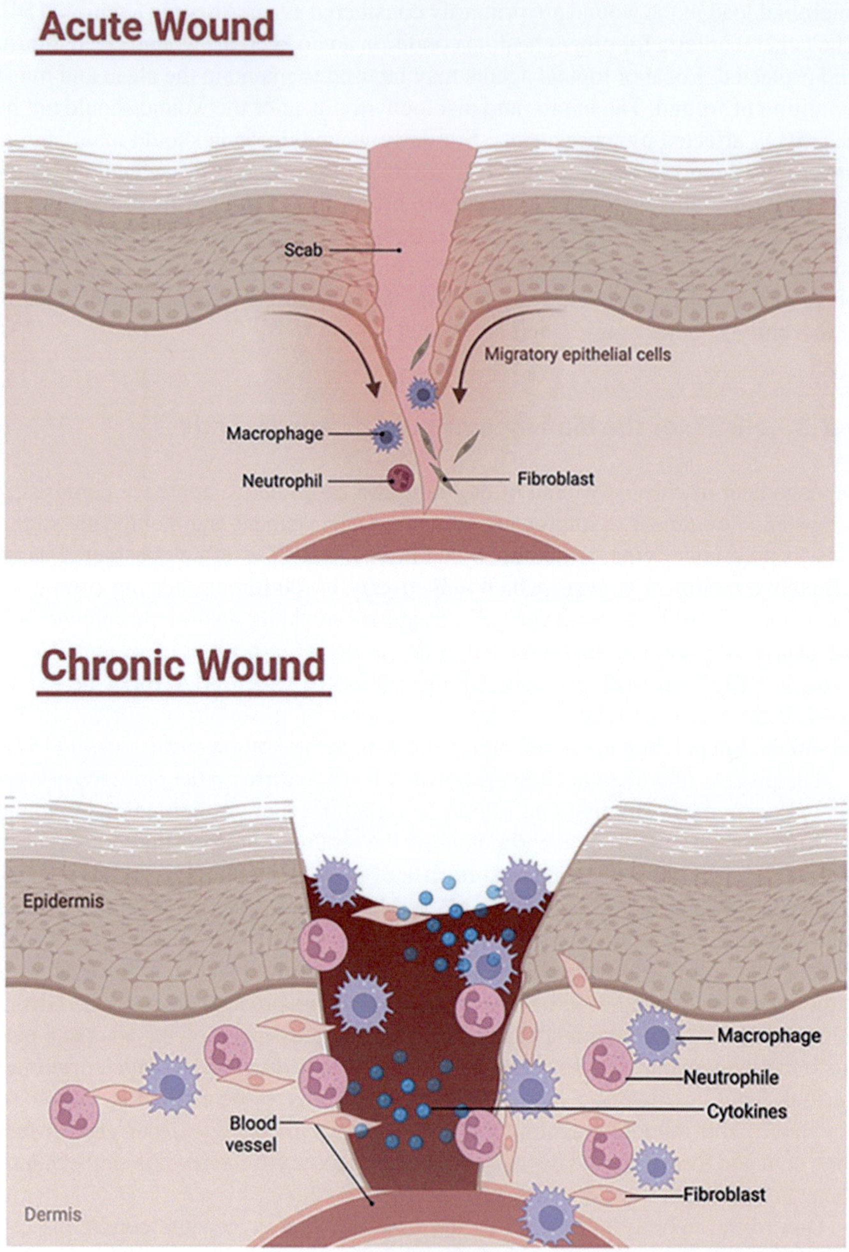

Fig. 7.11 Acute and chronic wounds. Acute wounds tend to heal through phases of normal healing leading to wound closure over time while a chronic wound shows delayed healing due to an elevated inflammatory response with poorly coordinated phases of wound healing

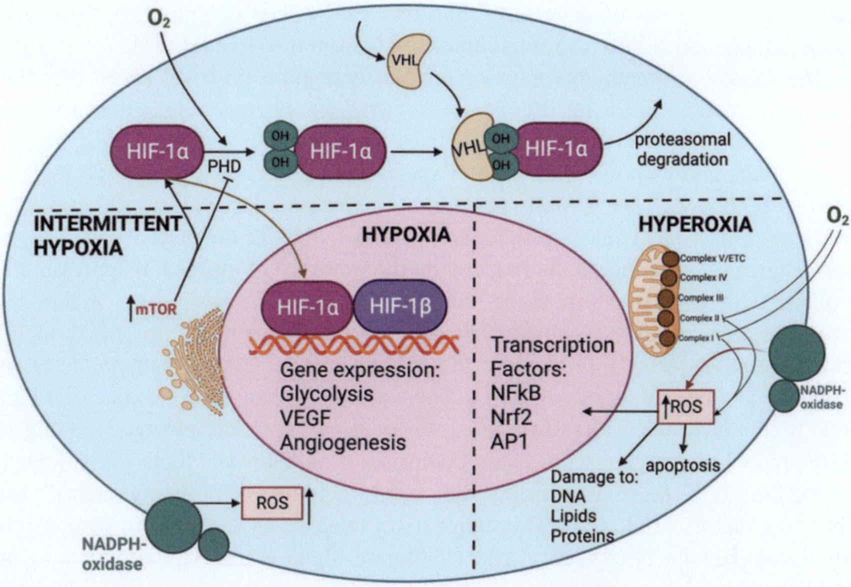

Fig. 7.12 Molecular mechanisms involved in cellular responses to various oxygen levels. Intermittent hypoxic conditions created during the course of HBOT result in the upregulation of genes involved in functions such as angiogenesis and energy transduction pathways. Additionally, anti-oxidant enzymes are activated in response to elevated ROS levels under hyperoxic cycle, which remains active long after the levels of ROS comes down during hypoxia

withdrawal from hyperbaric conditions and thus therapeutic effects are retained for longer period of time [158].

7.8 Gas Gangrene or Clostridial Myonecrosis

7.8.1 Etiology and Clinical Complications of Gas Gangrene

Gas gangrene or clostridial myonecrosis is a life-threatening infection of muscles mainly caused by anaerobic bacteria *Clostridium perfringens* and related species of clostridia. It is a progressive and highly invasive infection with the characteristics of extensive edema, cell necrosis, toxemia, and production of gas as waste product. Gas bubbles are formed in the blisters within the infected area as the infection progresses [159, 160]. Treatment of gas gangrene involves high doses of antibiotics and surgically removing the infected tissues. Clostridium, being an anaerobe, resides in the extensively damaged tissues and wounds which are poorly vascularized and have low oxygen levels. Soft tissue infection by clostridium may develop in hours but sometimes takes days to develop. Gas gangrene is characterized by the presence of swollen wound with bronze, gray or purple discoloration and putrid fluid discharge. Patients develop fever along with increased heart rate, rapid breathing, and suffer pain at the site of infection [161]. Gas gangrene is diagnosed by biopsy of tissue sample and culture examination of fluids from the wound [161].

The causative organism for gas gangrene is a gram positive, spore forming, encapsulated, non-motile bacilli of genus *clostridium.* There are more than 150 species of *clostridium*, of which some species are toxogenic such as *C perfringens, C. septicum,* and *C. novyi.* Some other species of *clostridium* are only proteolytic such as *C. histolyticum, C. bifermentans, C. sporogenes. C. fallax*, which do not cause gas gangrene [162]. *C. perfringens* is responsible for the majority of cases of gas gangrene. Gas gangrene is induced by the presence of clostridial spores in the wounds with extensive soft tissue damage and tissue necrosis. This makes the wounded area poorly oxygenated favoring the development of clostridial spores into vegetative form. Infection by clostridium species shows pathogenicity by releasing a wide variety of exotoxins. There are more than 20 exotoxins, of which nine toxins have been identified to cause local and systemic changes leading to development of gas gangrene. These comprises of alpha-toxin, theta-toxin, kappa-toxin, mu-toxin, nu-toxin, fibrinolysin, neuraminidase, "circulating factor" and "bursting factor" [163, 164]. The major toxin released by clostridium responsible for tissue damage is oxygen stable lecithinase alpha-toxin. Alpha-toxin acts on platelets and polymorphonuclear leucocytes leading to extensive capillary damage. All other toxins work together to facilitate the spreading of infection by causing damage too healthy tissue. The combined effect of all exotoxins leads to various clinical conditions such as hemolysis, jaundice, hemoglobinurea, tissue necrosis and exerts damage to organs systems in the form of renal failure, cardiotoxicity, and neurological dysfunction. The rate of toxin production by clostridium is high and spreads rapidly from the infected area to the healthy tissues. Anti-toxin is produced in response to released toxin but proves to be ineffective against rapid production of alpha-toxin and patient usually succumbs to death before active immunity can develop [164, 165].

The patients of gas gangrene requires extensive care and treatment as there is 100% mortality rate in patients suffering from gas gangrene if left untreated [166]. Even one out of four patients treated for gas gangrene succumbs to death. The treatment strategy against gas gangrene requires stabilizing the conditions of patient by providing blood transfusion to make up the hemolytic damage. High doses of antibiotics are required to curb further infection. Mostly penicillin is used for the treatment but other antibiotics may also be given in combination depending on other overlying infections [166, 167]. Also, clinical data suggests that using penicillin as a sole antibiotic does not affect toxin production and does not improve the survival rate in patients. Other antibiotics such as rifampicin, tetracycline, chloramphenicol, metronidazole, and clindamycin are targeted to inhibit protein production which aids in blocking the progress of infection by lowering alpha-toxin activity [168, 169]. Surgical removal of damaged and necrotic tissue is required as clostridium infections sets in rapidly in the presence of dead and decaying tissues.

7.8.2 HBOT as an Adjunctive Treatment for Gas Gangrene

HBOT was first employed as a treatment strategy against gas gangrene in 1960 [170]. Further clinical studies and research led to discovering the effectiveness of HBOT in improving the state of the patients with gas gangrene. Unsworth and Sharp reported an account of HBOT utilization for treatment of 73 gas gangrene patients over a period of 11 years in Australia and Papua New Guinea. As per the report, there were a total of seven deaths corresponding to 9.6% of mortality rate [171]. Similarly, Rudge reported a cumulative mortality rate of 23% in a review of 20 different clinical trials involving a total of 1200 gas gangrene patients treated with HBOT [172]. One of the important aspects of using HBOT is its ability to inhibit production of alpha-toxin. HBOT enables increase in the oxygen levels in the infected area and it has been reported that toxin production can be inhibited at 60 mm Hg or higher oxygen tension. However, the toxins already produced by the clostridium are not affected by HBOT [173, 174]. There are other additional benefits of HBOT which are also evident in its application for healing of non-infectious wounds as well. Hyperoxygenated state in the wound area leads to vasoconstriction reducing edema and allowing perfusion of ischemic and hypoxic tissues [174]. Relieving the hypoxic state diminishes clostridial growth and eliminates the acidic environment favorable for toxin production.

Hyperoxygenation also has a bactericidal effect as well as reduce the rate of germination of clostridial spores. HBOT promotes production of free radicals and ROS to induce bacterial killing and clearance [175]. Activation of immune cells and enhanced phagocytosis has also been reported under hyperbaric conditions in a large number of studies [127]. It should be noted that the necrotic dead tissues are not penetrated by oxygen or antibiotics due to the lack of any vasculature or connection. Such tissues need to be removed surgically prior to the therapy for better improvement. For the treatment of gas gangrene patient by HBOT, oxygen at 2.5–3.0 ATA pressure is sustained for 90 minutes with a break of 5 min after every 30 min. In the first 24 h of the treatment, three HBOT sessions are carried out, which is reduced to twice daily for the upcoming 2–5 days of treatment. The treatment is carried out until there is a clear demarcation of wounds and patients shows no sign of clinical toxicity (Fig. 7.13) [173, 176].

7.9 Necrotizing Soft Tissue Infections

7.9.1 Necrotizing Soft Tissue Infections—Complications and Treatment Strategies

Necrotizing soft tissue infections arise from ulcers or wounds affecting the deep fascia of skin and soft tissues (Fig. 7.14). It is characterized by the progression of infection into ischemic dermal necrosis involving the dermal blood vessels in the fascial layers [177]. There may be variations in the causative organism and site of infection but it is essentially a presence of necrotizing infection in one of the layers

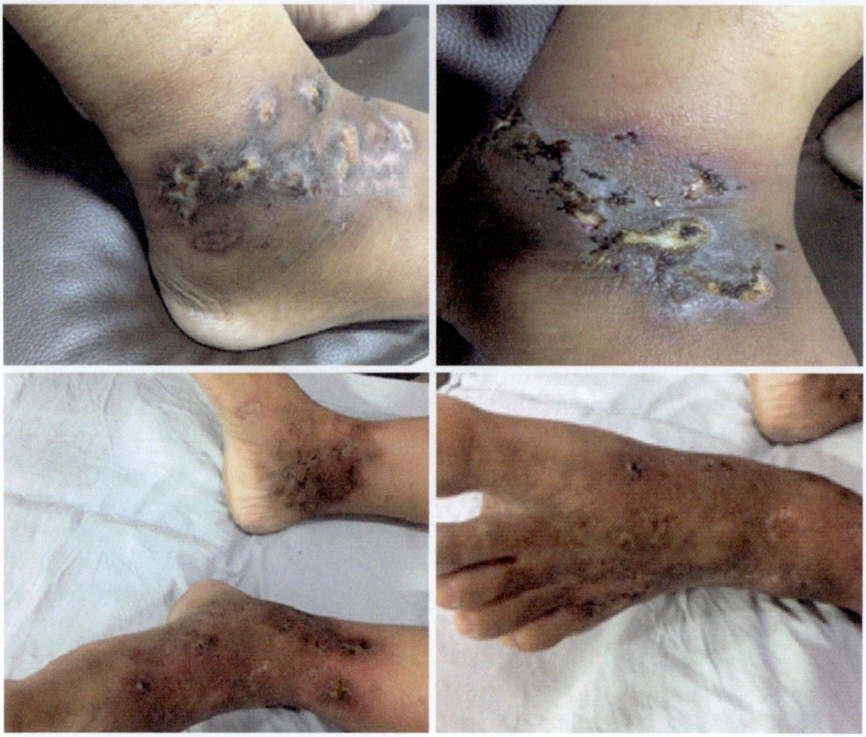

Fig. 7.13 Effect of HBOT on erythema gangrenosum

comprising the soft tissue, i.e., dermis, superficial fascia, deep fascia, muscles, or subcutaneous tissues [178]. The clinical manifestations of necrotizing soft tissue infection involve erythema, edema, fever and tachycardia. Pain is associated with the infection and hemorrhagic bullae, necrosis, sepsis, and subcutaneous production of gas develops as infection progresses. As the infection progresses from deep fascia to other planes of skin, dermal necrosis and other signs of spreading infection can be observed [179, 180].

On the basis of the initial site of infection, necrotizing soft tissue infection can be classified as myonecrosis and subcutaneous tissue infection. Subcutaneous tissue infection can be further distinguished as necrotizing fasciitis, involving infection of deep fascia, subcutaneous tissues, and superficial fascia and cellulitis, which involves superficial fascia and subcutaneous tissues but not deep fascial layer [181]. The common features of all these infections are its widespread nature and release of cloudy serous exudate. The wounded tissue is pale and necrotic with infiltration by leucocytes. Tissue necrosis occurs primarily due to extensive thrombosis in the small vessels of infected area. Due to the effect on smaller blood vessels, the wound area is poorly vascularized and hypoxic. The hypoxic conditions lead to lowering in oxidation-reduction potential and pH which in turn promotes growth of variety of strict and facultative anaerobic microbes. Local hypoxia within the infected area is

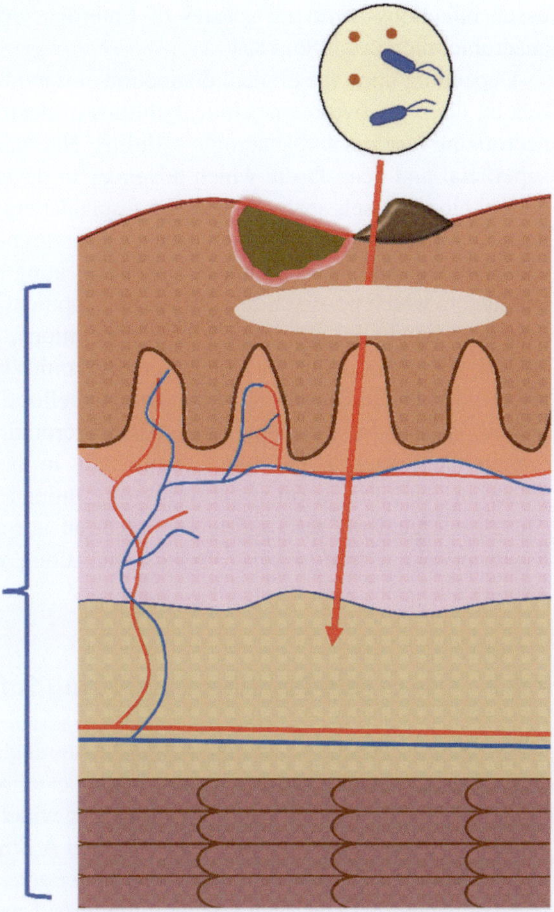

Fig. 7.14 Necrotizing soft tissue infection. Necrotizing soft tissue infections are the lethal infections affecting the layers comprising soft tissue compartment, i.e., dermis, subcutaneous tissues, superficial fascia or muscles

also associated with decreased perfusion and ischemia. Endothelial adherence of accumulated leucocytes affects immune function and substrate oxidation in the vicinity of wound. This leads to accumulation of hydrogen and methane gases within the tissues [181–183].

Necrotizing soft tissue infections are clinically classified on the basis of anatomical levels of dermal layers. However, as the infection progresses to later stages, it is difficult to differentiate these infections from each other. Microbiological examination of infectious species by microbial culture and gram stain provides more clear information about the type of necrotizing soft tissue infection [184, 185]. In most of the cases of necrotizing soft tissue infection, Group A, C or G beta-hemolytic *Streptococci* can be isolated from the infected tissue. Streptococcal infection is associated with infection from one or two other species of microbes in half of the cases. Infection from Group A beta-hemolytic *streptococci* releases toxins and cause ischemia, platelets, and neutrophil co-aggregation which affects blood flow to the infection site. Polymicrobial infection has been reported in most necrotizing soft tissue infections involving species of Enterobacteriaceae and various species of anaerobic microbes [186, 187].

Depending upon the clinical distinction and modes through which the infection sets in, there are two major clinical entities of necrotizing soft tissue infections—necrotizing fasciitis and anaerobic cellulitis. Necrotizing fasciitis is an infection of superficial and deep fascia which advances to develop into dermal necrosis after dermal blood vessels spanning through fascial layers are involved [188, 189]. It was initially termed as hemolytic streptococcal gangrene by Meleney in 1924 and the causative organism is often referred as "flesh eating bacteria" [190]. There are various factors which pose as a risk for the development of necrotizing fasciitis such as traumatic breaks in skin, insect bites, lacerations, deep abrasions, and puncture wounds. It occurs majorly in the lower extremities but any site in the body can be affected [191]. On the other hand, anaerobic cellulitis is the infection of the subcutaneous tissues and develops slowly than necrotizing fasciitis. The deeper fascia remains unaffected as the infection lies only in the superficial third upper layer. Anaerobic cellulitis is often caused during thoracic or abdominal surgeries. The causative organisms responsible for infection are a plethora of species such as *Staphylococci, Pseudomonas, Proteus,* etc. along with anaerobic non-hemolytic *Streptococci* [192, 193].

7.9.2 Use of HBOT against Necrotizing Soft Tissue Infections

Necrotizing soft tissue infections are life-threatening and need to be managed before its progression toward healthy tissues. Upon detection of infection, parenteral antibiotics are given to the patient. Penicillin G is a widely used antibiotic of choice due to its broad spectrum of action against Group A *Streptococci* and other anaerobic bacterial species such as *Clostridium and Fusobacterium.* However, with the prevalence of penicillin G resistant species, other broad spectrum antibiotics or combination of antibiotics are preferred such Rifampicin, Chloramphenicol, macrolides, and imidazole derivatives [194]. Surgical intervention involves the removal of necrotic tissues from the infection site in all the layers of soft tissues. Depending on the extent of infection, sometimes radical surgery involving amputation need to be carried out to check the spreading of infection and to save patients' life [195, 196].

Based on the large number of experimental and clinical studies, the jury of European committee for hyperbaric medicine (ECHM) consensus conference in 1994 and 2004 recommended HBOT against anaerobic soft tissue infection [197]. HBOT employs the effects of oxygen in hyperbaric conditions to fight against the infection. The rationale for the use of HBOT relies on the direct effect of oxygen on anaerobic bacteria and its beneficial properties to enhance the natural immune system and facilitating the action of antibiotics [198]. HBOT is used as an adjunctive therapy to standard wound care regime. Thus, debridement of necrotic tissues, removal of abscess and administration of antibiotics is continued during the course of HBOT. Additionally, antibiotics which blocks protein synthesis such as clindamycin and linezolid are used to inhibit the production of exotoxins [199]. HBOT has proved to be highly effective in improving the condition in necrotizing soft tissue infections (Fig. 7.15). HBOT is administered at 2-2-5 ATA pressure for 90 minutes using 100% oxygen twice a day for initial few days until no signs of necrosis are observed. In case of diagnosis of clostridial myonecrosis, HBOT is administered at 2.8–3.0 ATA pressure following the protocol of three sessions in first 24 hours until the patient is stable. In order to monitor the effects of HBOT, partial pressure of oxygen in the infected area and bacterial clearance from the wound are regularly monitored [200, 201].

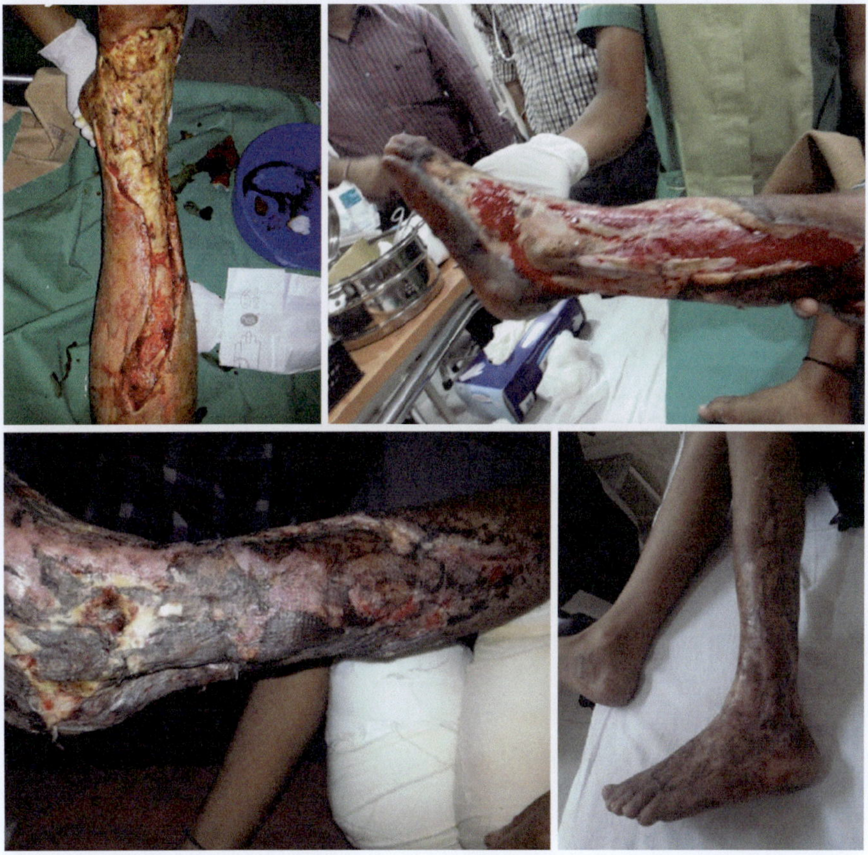

Fig. 7.15 Effect of HBOT on necrotizing fasciitis

7.10 Refractory Osteomyelitis

7.10.1 Refractory Osteomyelitis

Osteomyelitis is an inflammatory condition of bone which arises due to infection from bacteria or mycobacteria. There is an occurrence of ischemia and hypoxia in the affected region of bone tissue and the condition could be acute, subacute or chronic [202, 203]. Chronic osteomyelitis is defined as a bone infection that persists for a period of over 6 months with histological and radiological evidences of infection. Refractory osteomyelitis is the case of chronic osteomyelitis which fails to heal and persists even after surgical, antibiotic or other appropriate interventions (Fig. 7.16). The consequences of refractory osteomyelitis may be severe and the limb salvage may not always be possible [204]. Management of osteomyelitis and patient care requires weeks of hospitalization and subsequent treatment plan. The infection from bones has a tendency to spread into adjacent tissues and organs resulting in multiple organ failure and death [205]. Presence of predisposing conditions such as diabetes mellitus and peripheral vascular disease increases the risk of contracting osteomyelitis [206]. In many cases patients with chronic non-healing wounds may develop underlying osteomyelitis due to the presence of persistent wound.

Infection may arise in osteomyelitis as a result of hematogenous spread or direct infection of bones. The most common organisms responsible for hematogenous spread in infants and children are Staphylococcus and Streptococcus while *S. aureus* is commonly found in case of adults [204, 207]. Chronic refractory osteomyelitis develops as the infectious condition shows no response to culture directed antibiotic treatment and surgical debridement. Although 70–80% of primary osteomyelitis cases are cured by surgical intervention and antibiotic therapy, there is still 20–30%

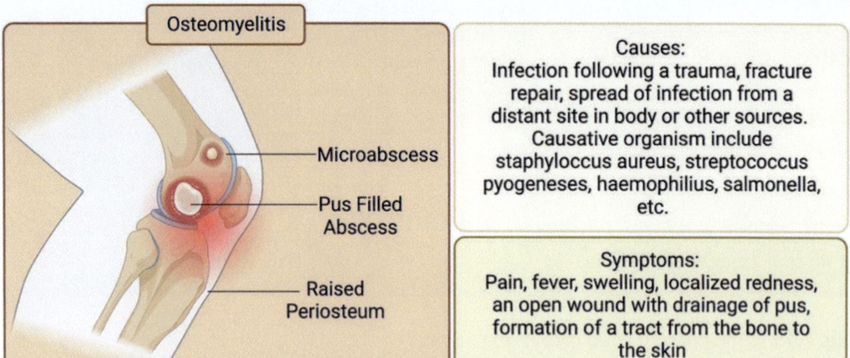

Fig. 7.16 Cause and symptoms of refractory osteomyelitis. Refractory osteomyelitis is characterized by infection that occur after acute osteomyelitis fails to respond to ongoing treatment regime. The infection may be caused after trauma or fracture leading to the formation of abscess in the affected bone tissue

recurrence rate where treatment strategies were not effective. Patients, that are immunocompromised, have pre-existing conditions, inadequate debridement, poor vasculature or poor soft tissue reconstruction can benefit from HBOT as an adjunctive treatment [208–210].

7.10.2 Clinical Management of Refractory Osteomyelitis

Surgical intervention involves draining of abscess, foreign body removal, debridement of sequestrum and enabling normal vascularization and obliteration of dead spaces formed due to debridement. There may be requirement of radical resection of bones, disarticulation of joints or amputation of affected extremity [211]. Apart from these, appropriate antibiotic course is required to check the growth of microbes after surgery. If there is an absence of infection after the initial treatment, reconstructive surgery and soft tissue reconstruction can be planned. However, recurrence rates are still high even after surgical intervention and antibiotic treatment demanding the need of adjunctive therapeutic strategies such as HBOT. There are no randomized clinical trials to study the effects of HBOT against refractory osteomyelitis. However, non-randomized trials and studies on animal models suggest the effectiveness of HBOT in improving the condition in refractory osteomyelitis. Clinical studies have shown that bone oxygen tension is reduced in osteomyelitis [2, 209, 210]. Thus, it is beneficial to elevate the oxygen content in affected bone to facilitate the microbial killing as well as to improve the antibiotics availability and stimulation of osteogenesis.

7.10.3 Cierny-Mader Classification of Osteomyelitis

Different stages of osteomyelitis and advice to utilize HBOT as an adjunctive treatment has been elaborated in Cierny-Mader classification system (Fig. 7.17; Table 7.2). It gives an account of patient classification as normal host, compromised

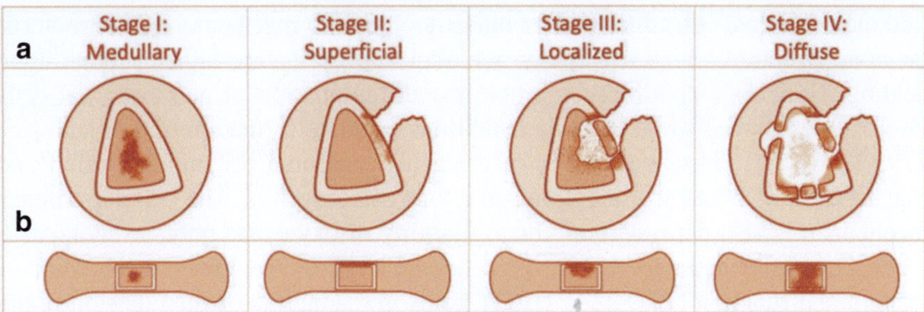

Fig. 7.17 Cierny-Mader classification of osteomyelitis

Table 7.2 Cierny-Mader classification of chronic osteomyelitis

Anatomical type		Physiological classification	
Stage 1	Medullary osteomyelitis	A Host	Normal Host
Stage 2	Superficial osteomyelitis	BS Host	Systemic compromise
Stage 3	Localized osteomyelitis	BL Host	Local compromise
Stage 4	Diffuse osteomyelitis	BSL Host	Systemic and Local compromise
		C Host	Treatment worse than disease

host or patients for whom treatment is worse than disease [212]. Cierny-Mader classification of chronic osteomyelitis patients is based on their anatomical and physiological parameters. The patients in stages 3B and 4B are considered ideal candidates for HBOT. Osteomyelitic patients with pre-existing conditions such as diabetes mellitus, peripheral vascular disease or extensive soft tissue scarring can have potential benefit from HBOT. Similarly, improved outcomes are expected in case of immunocompromised patients as well as patients with poor or inadequate debridement, poor soft tissue reconstruction and poor vascularity [213, 214].

7.10.4 Treatment of Refractory Osteomyelitis by HBOT

Pathophysiology of refractory osteomyelitis demonstrates local edema, necrosis, and hypoxia with decreased blood flow. The hypoxic condition promotes the infectious state of osteomyelitic bone [215, 216]. Intramedullary partial pressure of oxygen below 30 mm Hg gives an indication of osteomyelitis [217]. Hypoxic conditions arise primarily due to three main factors—oxygen consumption by microbes, diminished local perfusion due to edema, and oxygen consumption by inflammatory cells. Oxygen is a critical factor for healing of wounds as it is involved in multiple functions related to immunological functions and tissue regeneration. Reversal of hypoxic conditions into hyperoxic conditions by administration of oxygen under hyperbaric conditions aids in relieving the pathological state in refractory osteomyelitis [218]. Under hyperoxic conditions, there is an increased mobilization of fibroblasts at the site of injury. Fibroblasts produce collagen in an oxygen dependent manner, which deposits as the extracellular matrix in the healing wound. Fibroblasts-like mesenchymal cells differentiate into osteoblasts to regenerate immature fibrillar bone structure which eventually gets replaced by mature lamellar bone upon healing. Hyperoxic conditions promote the differentiation of mesenchymal cells into osseous tissues while hypoxic conditions promote formation of cartilage [23, 219, 220, 125]. Thus, availability of oxygen is essential for mineralization and regeneration of bones to heal from refractory osteomyelitis. The cellular defense mechanism is also dependent on the availability of oxygen as polymorphonuclear leucocytes and phagocytes require oxygen to generate oxygen free radicals for their bactericidal function [221, 222]. HBOT helps in raising the oxygen tension in the ischemic and hypoxic tissues enhancing the neovascularization around the healthy bone tissues. Neovascularization enables infiltration of immune cells, antibodies, and antibiotics to the affected area to facilitate healing [223]. The osteoclast activity is enhanced in the presence of oxygen and helps to remove the bony debris [224]. The recommended course of HBOT for the treatment of refractory osteomyelitis is administration of 2–3 ATA of 100% oxygen for 90–120 minutes twice or thrice until the conditions are improved [2, 209, 210].

Various reports are available on the improved conditions in refractory osteomyelitis upon administration of HBOT. HBOT is an important therapeutic intervention especially in case of post-traumatic osteomyelitis due to bone fracture (Fig. 7.18). In order to acquire maximum benefit from HBOT, it is essential to follow an aggressive surgical regime, bone debridement, and antibiotic treatment. Osteomyelitis adjacent to the central nervous system or other vital organs requires extensive surgical debridement and antibiotic therapy prior to the administration of HBOT as it significantly affects the morbidity and mortality if the infection spreads to the sensitive regions. The course of antibiotics and HBOT are usually continued until the surgically debrided bones are properly revascularized and there are no signs of infection.

7.11 Malignant Otitis Externa

7.11.1 Malignant Otitis Externa

Malignant otitis externa (MOE) is an infection of external auditory canal with higher susceptibility in immunosuppressed individuals and patients with diabetes mellitus [225]. The major causative organism responsible for the infection is *Pseudomonas aeruginosa* which can be isolated from the aural discharge. The origin of the infection can also be fungal with *Aspergillus fumigatus* being the major causative organism [226, 227]. The infection may result in the osteomyelitis of the temporal bone which can be invasive. MOE may lead to other serious complications involving nerves which could have fatal consequences. It is reported in all age groups but is found to be more common in older patients. The course of treatment

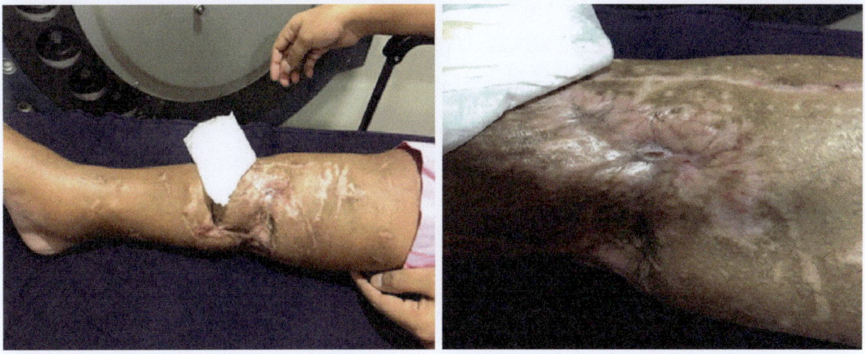

Fig. 7.18 Effect of HBOT on osteomyelitis

for MOE consists of surgical debridement and extensive antibiotic intervention [228, 229]. Anti-pseudomonal medicines such as quinolone can be administered as an oral and topical agent [230]. The mortality rate from MOE has been reduced from 50% to 10–20% with the administration of better and aggressive treatment regimens [231, 232].

The infection in MOE causes bone erosions involving the base of skull through fascial planes and venous sinuses and can spread to surrounding tissues toward cranial nerves and intracranial region (Fig. 7.19) [233]. The biopsy of external auditory canal shows ulceration and epithelium loss due to infection. Inflammation is evident in the dense fibrous tissues with reactive changes in the form of mild to prominent hyperplasia [234]. The major symptoms of MOE involve pain in the auditory canal, edema, abscess, and fluid discharge from ear. Imaging techniques such as CT scan and magnetic resonance imaging along with tissue sampling and culture are utilized for the diagnosis of MOE [235].

7.11.2 Myringotomy as an Assessment Tool in Malignant Otitis Externa

Myringotomy is a surgical procedure which involves making a small incision in tympanic membrane of ear. The incision is made in order to access the middle ear and drain fluid in case of otitis media with effusion. Fluid drainage from middle ear is carried out by placing a tube called myringotomy tube into the incision [236].

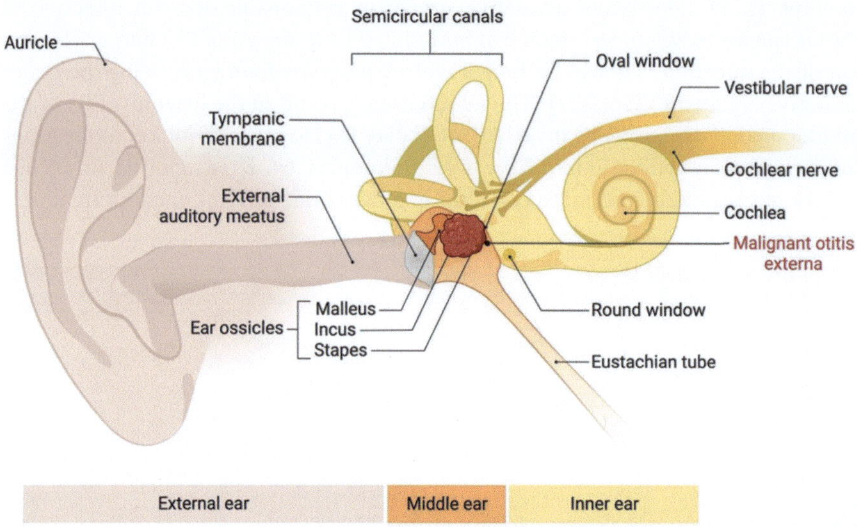

Fig. 7.19 Malignant otitis externa. Malignant otitis externa is an aggressive infection of external auditory canal and temporal bone. The infection is often initiated by iatrogenic trauma or self-inflicted injury to the external auditory canal. There are clinical findings of granular tissues in the auditory canal at the bone-cartilage junction

Otoscopy and Tympanogram is performed before the procedure of Myringotomy to determine the response of ear drum toward the changes in pressure [237]. Otitis media is an infection of middle ear that may occur as an extended infection of otitis externa and considered as a potential reason for hematogenous spread [238]. The extent and severity of infection is determined by diagnostic imaging techniques such as CT scan, MRI or radiography. Drainage of fluid from ear is required prior to the treatment of infection and administration of adjunctive therapies like HBOT. Myringotomy helps in relieving the pain and pressure built in the ear along with reducing the chances of infection to spread further.

7.11.3 Management of Malignant Otitis Externa by HBOT

The patients of MOE are extensively administered with antibiotics for 4–6 weeks to check the spread of infection. HBOT has been shown to exert anti-microbial effect on *P. aeruginosa* and thus recommended as an adjunctive therapy for the treatment of MOE [239]. There are no control studies regarding the use of HBOT in the treatment of MOE. However, there are several case reports advocating its efficacy against MOE. Bath et al. successfully treated a case of MOE with the involvement of optic nerve [240]. Similar results were obtained by Gilain et al. for a case of MOE involving facial nerve [241]. Shupak et al. worked with two patients of MOE involving infection in the face and skull [242]. There are several other reports involving a larger number of patients who benefitted from HBOT. Davis et al. treated 16 patients of MOE successfully using HBOT out of which 6 were presented with severe complications [243]. Similarly, in two separate cases, Martel et al. and Tisch et al. utilize HBOT as an adjunctive treatment along with antibiotics and immunoglobulins, respectively, for the treatment of 22 patients with MOE [244, 245]. Overall, it can be stated that HBOT functions as a useful technique against MOE due to its well established anti-microbial activity. The recommended protocol for utilizing HBOT in the treatment of MOE involves administration of 100% oxygen at 1.5 ATA for 60–90 min [246, 247]. The HBOT sessions can be carried out for up to 8 weeks if required to limit the chances of recurrence and spreading of infection to the nervous system.

7.12 Compromised Skin Grafts and Flaps

7.12.1 Considerations for Skin Grafts and Flaps

Skin grafts and flaps are indispensable for the management of wounds requiring reconstructive surgery [248, 249]. In case of grafts and flaps, wound healing is a complex process as it requires overcoming ischemic and hypoxic conditions in order to heal. Adequate perfusion and oxygenation of tissues to the local environment of wound are critical factors for the success of graft [250]. Apart from these, antibiotics and suitable wound dressing are required for the proper management of wound. Similar to the healing of a normal wound, skin grafts and flaps goes through the process of granulation, epithelization, and contraction [251]. Reconstructive surgeries often involve transfer of skin grafts or flaps derived from healthy body part to the affected area (Fig. 7.20). Survival of skin grafts or flaps depends upon multiple factors and failure of any stage of healing process reduces the chances of graft survival. Production of free radicals by the neutrophils plays an important role in the ischemic damage caused to the graft tissues. There is a constriction of lumen in the microvasculature due to edema and necrosis of muscle cells and endothelial lining of blood vessels. Level of creatine kinase in the plasma is considered as an indicator of graft survival since its levels are significantly higher in case of graft failure [252, 253]. Thus, in order for successful establishment of skin grafts and flaps several factors such as free radical damage, ischemia and tissue hypoxia needs to be taken into consideration.

7.12.2 Wound Management on Compromised Grafts and Flaps

Skin grafts consist of a segment of skin to be transferred to a new site in the body. It is primarily composed of dermis and epidermis without any vascular support. The viability of skin graft in the transplant site depends on the oxygen diffusion from the wound bed until new vasculature develops via angiogenesis [254]. During the initial period after transplant, there is low level of oxygen in the skin graft. Thus, it is critical to provide optimum conditions in order to establish the skin graft in the recipient bed. Skin grafts can be taken from the donor site as a split thickness skin graft

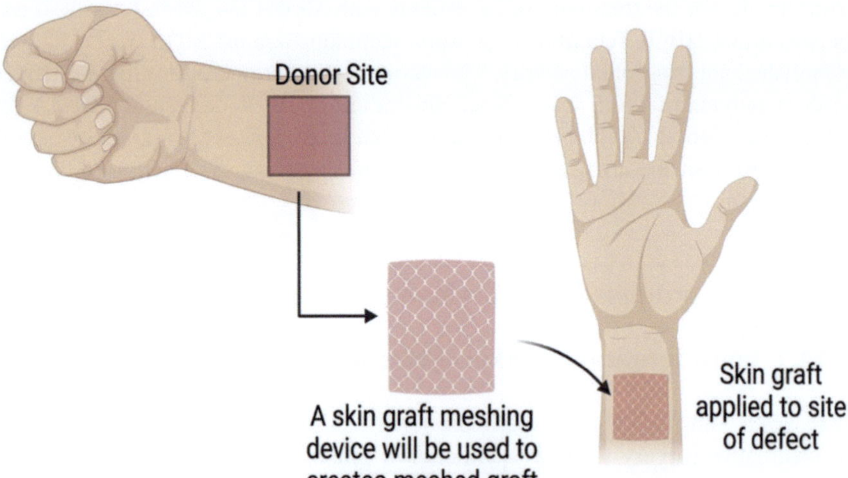

Fig. 7.20 Procedure of skin grafts and flaps. Skin grafts and flaps are removed from the donor site and transferred to the recipient site in the body. Skin grafts are devoid of any vascular support while flaps have their blood supply intact via a pedicle

(consisting of entire epidermis and apart of dermis) or full thickness skin graft (consisting of entire epidermis and dermis). On the basis of tissue composition, grafts can be classified as simple or composite grafts. Simple grafts consist of a single type of tissue while a composite graft is composed of more than one tissue type along with subcutaneous fat, full thickness skin and cartilage. The conditions for survival of a composite graft are more complex than for a simple graft. Based on the origin of the donor, grafts can be defined as autografts (derived from self), allografts (derived from a different person) or xenografts (derived from another species) [251, 255–257].

Flaps differ from grafts in having intact vasculature along with tissues. Flaps are more favored in reconstructive surgeries where the recipient bed has a poor or impaired vasculature [249]. Flaps can be distinguished by the site of their removal with respect to the recipient site as local flaps (derived from the tissue in proximity to the recipient site), regional flaps (derived from the tissues from the same general area of recipient site), and distant flaps (derived from the area away from the recipient site). On the basis of blood supply, flaps can be classified as random, axial, pedicle or free [258–260].

In case of both skin grafts and flaps, ischemia is the major cause for transplant failure. Freshly incorporated grafts or flaps are poorly oxygenated leading to inadequate metabolic functions. The hypoxic condition, in general, interferes with proper wound healing. Ischemia-reperfusion injury occurs under hypoxic conditions and increased production of free radicals leading to lipid peroxidation and adherence of leucocytes to the blood vessels [261]. Skin grafts and flaps get compromised due to occlusion in the vessels which restricts the passage of blood cells. Prolonged ischemic conditions result in tissue necrosis due to ischemia-reperfusion injury and failure to re-establish blood flow to the transplanted graft or flap [262]. There are other factors such as age of the patient, other elative therapy, smoking or systemic diseases that may interfere with the proper healing of wound after graft or flap transplant. Change in appearance of graft tissue is indicative of viability of graft and the progress in the re-establishment of graft vasculature [263, 264]. Methods like transcutaneous oxygen pressure, tissue pH, laser Doppler and radioisotopic analysis are employed for the prognosis of transplanted skin graft and flaps [265, 266]. The chances of survival of skin grafts and flaps can be greatly improved by administration of adjunctive strategies such as vasodilators, free radical scavengers, fibrinolytic agents, dextranes, and HBOT [267–271].

7.12.3 HBOT for Treatment of Compromised Grafts and Flaps

Administration of HBOT as an adjunct to reconstructive surgery reduces the failure rate to 4.5% from the usual rate of 10% [272]. Although, HBOT is not as widely used as a modality in reconstructive surgeries; its utilization is known to greatly improve the outcomes. HBOT is recommended when there is any doubt in the viability of graft or flap due to ischemia-reperfusion injury. HBOT helps in resolving the tissue edema and re-establishment of proper blood flow to the graft [273]. The patients are administered with HBOT after surgery at 2–2.5 ATA of 100% oxygen for 120 min for 3 days [274]. HBOT has been studied as an adjunctive therapy in a group of 24 patients where 92% surface area of graft was observed to survive in comparison to 63% in the control group. In the same study overall 65% of patients experienced completely successful graft as compared to 17% of patients in the control group [267]. Graft tissue is hypoxic in the early stages of transplant with low rate of survival. Administration of HBOT increases the tissue oxygen tension in the dermis and elevates the fraction of diffused oxygen to the graft from the recipient bed [275]. Also, the donor site shows accelerated healing under hyperbaric conditions. The role of HBOT in the revascularization is not fully understood yet but reattachment by microvascular anastomosis has been reported by the use of HBOT [276, 277]. Diagnosis of compromised graft and its response to HBOT need to be carefully assessed by the surgeon and hyperbaric physician to improve the chances of successful graft.

7.13 Crush Injuries

7.13.1 Classification of Crush Injuries and Challenges

Crush injuries are defined as a spectrum of mechanical injuries to the body as a result of being subjected to high degree of force or pressure. The impact of injury is severe affecting two or more tissue types (muscles, bones, nerves, etc.) [278]. The major concern with crush injuries is to retain the viability of the injured tissues and even if it is possible to salvage the tissues, patient may suffer with functional deficits. The presentation of injuries ranges from minor contusions to severe irreversible damage, requiring limb amputation. There is often severe damage to soft tissues along with bones, nerves, and vascular structures. The extent of damage to the tissues depends on the impact of traumatic force and can manifest as open wound injury, fracture, severe bleeding, and compartment syndrome [279, 280].

The complications arise in case of crush injuries due to the involvement of multiple types of tissues. There is a high risk of infection and impaired tissue healing which may lead to loss of function in the affected limb. Maintaining the viability of tissues after crush injury is the major immediate challenge since the vascular damage and blood loss results in ischemic conditions [281, 282]. There is a decreased supply of blood and thrombosis affecting microvasculature of the affected region. Thus, it is essential to take measures in order to salvage injured but still viable tissues before necrotic conditions sets in. In case of mild injuries, there are less complications and the condition can be managed by minimal medical interventions. However, severe injuries pose greater threats as they are characterized by soft tissue loss, compound fractures, arterial injuries and marginally viable tissues. There is a risk of irreversible damage and several delayed complications may also occur in the form of osteomyelitis, non-joining fractures, and requirement of delayed amputation [283, 284]. Crush injuries are classified primarily by using Gustilo's classification system of open fractures which helps the physician to plan their treatment strategy and management of patient's condition [285, 286] (Table 7.3).

7.13.2 Physiological and Biochemical Aspects of Crush Injuries

Tissues affected by crush injuries undergo edema and ischemia which may lead to the necrosis of partially viable adjacent tissues. Extensive vascular damage and edema generates tissue hypoxia which in turn impairs the process of wound healing and tissue regeneration. Hypoxia originates partially as a result of blood cell clumping at the site of injury interfering with the microcirculation [287]. Cellular hypoxia also leads to the downregulation of cellular metabolism further reducing their viability. The hypoxic conditions compromise immune functions as well as different stages of wound healing such as collagen production and neovascularization [288]. Oxygen supply to the tissues within the site of injury is largely dependent upon plasma dissolved oxygen but it is not sufficient to fulfill the metabolic demand of cells. Other complications may also arise due to severity of crush injuries. The fluid pressure gets elevated in a closed compartment upon injury which impairs perfusion within the compartment and creates ischemic conditions. The extent of tissue damage is analyzed by monitoring the value of serum myoglobin, potassium, BUN (blood urea nitrogen), and creatinine [289]. In order to assess the condition of patient and top categorize the crush injury various methods such as measurement of compartment fluid pressure, radiography to analyze fractures and angiography to determine the neurovascular damage are employed [290, 291].

7.13.3 Effect of HBOT on Wound Healing in Crush Injuries

Crush injuries require immediate medical intervention in order to prevent permanent damage to the affected region and to stabilize the patient's condition. In severe

Table 7.3 Gustilo classification of crush injuries

Type	Mechanism	Expected outcome	Infection rate	Amputation rate
I	Small laceration <1 cm	Not different from a closed fracture	Minimal	
II	Large laceration with minimal soft tissue damage	Not different from a closed fracture	3%	
III	Crush Injuries			
A	Sufficient soft tissue to close wound (primary or delayed)	<10% complication rate	4%	0%
B	Requirement of flaps or grafts to cover bone	>50% complication rate	52%	16%
C	Major vessel injury	About 50% complication rate	42%	42%

cases of crush injuries, the risk of complications is high and may require multiple surgeries [292, 293]. The primary goal of crush injury treatment is restoring blood circulation to avoid tissue ischemia and ischemia associated secondary complications. Apart from the usual treatment regimens for crush injuries, HBOT is considered as a useful adjunct treatment (Fig. 7.21). HBOT enables several functions critical for wound healing which promotes body's ability to heal effectively [294, 295]. The typical HBOT protocol for treatment of crush injury requires 2 ATA of 100% oxygen for 90 minutes. The treatment is initiated within first 24 hours of primary treatment followed by two sessions daily for 3 days [296]. The duration of treatment can be extended based on the clinical judgment of physician so as to manage wound healing and support reparative surgeries.

HBOT results in the hyperoxygenation of plasma promoting its availability to the hypoxic margins of wound. This helps against tissue hypoxia and ischemia, thus keeping the tissues viable during the treatment [288]. There are various other secondary effects of hyperoxygenation which enables effective healing of wound. Vasoconstriction occurs under hyperbaric conditions resulting in reduced edema. Oxygen dependent processes such as fibroblast functions, collagen production,

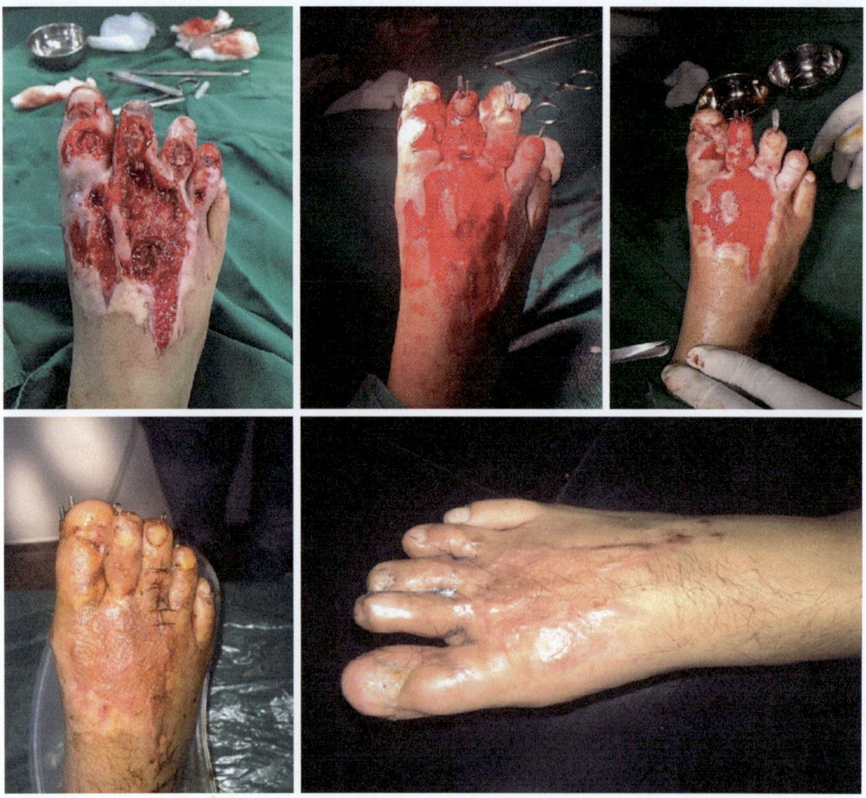

Fig. 7.21 Effect of HBOT on crush injury

neutrophil mediated anti-microbial action, and osteogenesis are enhanced promoting effective wound healing [294, 297, 298]. Under high oxygen tension, the process of angiogenesis is triggered which aids in restoring blood supply to the viable tissues. It is evident in various reports that the use of HBOT in crush injuries improves the status of wound healing and reduces the probability of taking drastic measures in the form of limb amputation [295].

7.14 Severe Anemia

7.14.1 Clinical Scenario of Severe Blood Loss and Complications

Hemoglobin (Hb) is the oxygen carrying pigment in the red blood cells which distributes oxygen to all organ systems through micro-capillary beds. In case of a marked decrease in the red blood cell mass due to hemorrhage, hemolysis or aplasia, there is an insufficient oxygen supply to the tissues leading to development of anemia [299, 300]. The symptoms of anemia are manifested as feeling of dizziness, chest pain, weakness, increased heart rate, and shortness of breath (Fig. 7.22). Hb can carry 1.38 ml of oxygen per gram while approximately 0.003 ml of oxygen is carried in dissolved form in 1 ml of plasma at normal atmospheric pressure [301]. The overall arterial oxygen content (CaO_2) can be determined by using the following formula:

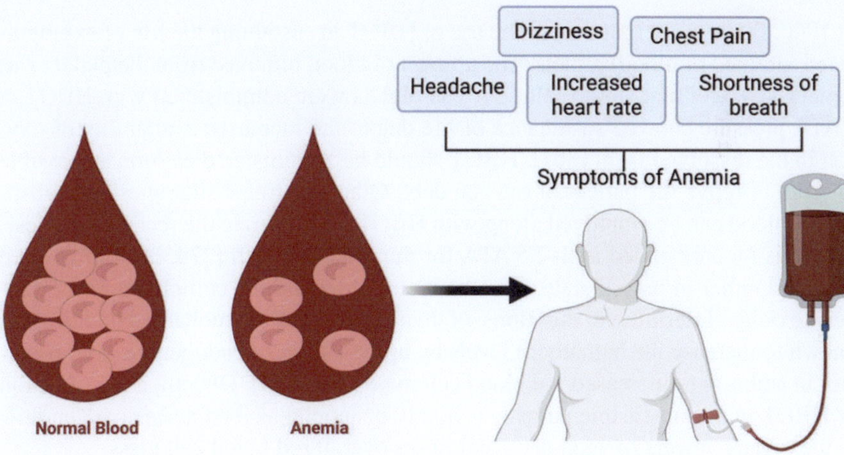

Fig. 7.22 Symptoms of anemia. Anemia is a condition of inadequate red blood cells in the blood to ensure optimum transport of oxygen to the tissues. As a result of improper oxygen supply to the tissues, symptoms such as dizziness, chest pain, headache increased heart rate, and shortness of breath are experienced by the anemic patients

$$CaO_2 = [\text{Hb level in blood}(\text{gm/dL})] \times 1.38\,\text{ml}\,O_2) + [0.003\,\text{ml}\,O_2 \times paO_2\,(\text{mm Hg})]$$

Under normal conditions, 5–6 ml of oxygen is carried by 100 ml of blood which is distributed to the entire body through the network of microvasculature. The normal Hb content of a healthy individual is 12–16 gm/dl [302]. The oxygen delivery mechanism becomes inadequate as the Hb level in the blood drops down to a critical value (below 5 gm/dl). Severe anemic conditions arise in case of exceptional blood loss. This creates a condition of cellular hypoxia due to insufficient supply of blood to the tissues. As a result, there is an accumulation of oxygen debt in anemic patients [303, 304]. The metabolic demand in hypoxic conditions is not fulfilled by aerobic respiration. This leads to the initiation of anaerobic respiration in the form of lactic acid pathway [305]. Thus, there is a deficit of oxygen required to remove the accumulated lactic acid in the tissues. Blood transfusion is the frequently used measure to overcome oxygen debt and to relieve anemic conditions [306]. However, blood transfusion may not be always possible due to immunological conditions or non-availability of required blood. Such patients need to be administered alternative treatment to alleviate the complications of severe anemia. HBOT has been found to be effective in overcoming the situation of accumulating oxygen debt and highly beneficial for the acute anemic patients.

7.14.2 Management of Severe Anemia by HBOT

In 1960, Boerema demonstrated the use of HBOT in sustaining the life of exsanguinated (drained of blood) piglets. The amount of blood removed from the piglets was replaced by acellular saline solution. The piglets were administered with HBOT at 3 ATA pressure showed sustenance of life due to the increased availability of oxygen in the dissolved form [307]. HBOT should be administered as soon as possible in order to relive the patient of oxygen debt. Other treatment strategies and substitute of blood can be employed along with HBOT to accelerate the recovery process. HBOT is recommended at 2–2.5 ATA for duration of 3–4 h [274, 308]. It is often provided either in pulsed state or intermittently to avoid the complications of oxygen toxicity. Intermittent durations of hyperbaric and normobaric conditions is known to increase the hematocrit levels by upregulation of endogenous erythropoietin in addition to increased red blood cell mass [308, 309]. Overall, administration of HBOT in untransfusible patients is highly beneficial as it enhances oxygen flow to the organs, lowers oxygen debt, and raises overall red blood cell mass.

7.15 Sickle Cell Anemia

7.15.1 Pathophysiology and Clinical Aspects of Sickle Cell Anemia

Sickle cell disease comprises of a group of autosomal recessive diseases marked by the high concentration of hemoglobin S (HbS) in the blood instead of the normal form hemoglobin A (HbA). Sickle cell anemia is the most common form of sickle cell disease which affects the individuals homozygous for the abnormal β-globin gene at chromosome 11 while the individuals carrying one copy of this gene are carriers of the disease. The abnormal HbS exists due to a single nucleotide polymorphism (SNP) in the β-globin gene, which cause a substitution of glutamate residue with valine residue at position 6 of β-globin polypeptide chain (Fig. 7.23). Under normal condition, the mutation is benign and does not affect the structural integrity of Hb. However, under hypoxic conditions, polymerization of HbS takes place due to exposure of hydrophobic patches resulting in clumping of blood cells. The blood cells get deformed in the shape of a sickle obstructing the normal flow of blood [310–314].

The pathology of sickle cell anemia arises primarily due to the loss of elasticity of red blood cells. Red blood cells (RBCs) assume abnormal sickle shape under conditions of low oxygen availability. Normal RBCs are biconcave disc shaped and flexible to pass through fine capillaries while the sickle shaped RBCs are rigid and get stuck in the narrow capillaries obstructing the blood flow. These abnormal RBCs are short lived as compared to normal blood cells and rapidly undergo hemolysis due to damage to the cellular membrane [310, 315–317]. This creates anemic

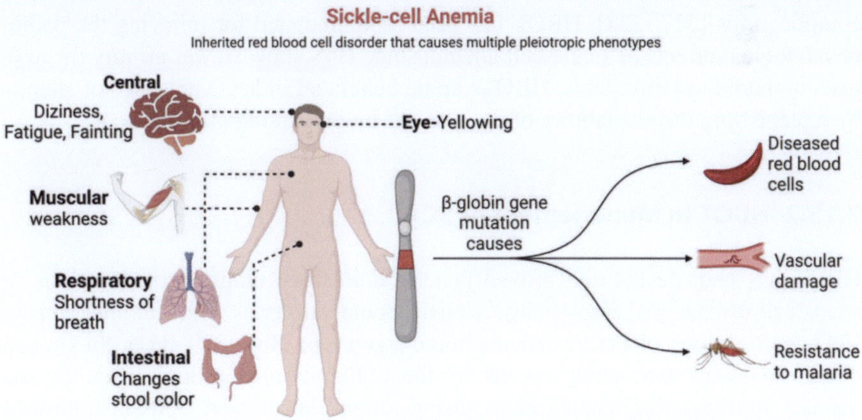

Fig. 7.23 Sickle cell disease. Anemia is a condition of inadequate red blood cells in the blood to ensure optimum transport of oxygen to the tissues. As a result of improper oxygen supply to the tissues, symptoms such as dizziness, chest pain, headache increased heart rate, and shortness of breath are experienced by the anemic patients. Conversely, sickle cell trait provides resistance from malaria

conditions affecting the oxygen carrying capacity of blood and generating ischemic conditions. The rate of destruction of RBCs is usually higher than the production of new blood cells in the bone marrow. There are three independent factors affecting the rate of HbS polymerization—Hb concentration, presence or absence of fetal Hb and degree of intracellular deoxygenation. Cell clumping in the microvasculature is promoted by other related factors induced by HbS [318]. The amount of circulatory endothelial cells is increased along with the increased expression of adhesion molecules both on endothelial cells (VCAM1 and CD36) and RBCs ($\alpha 4\beta 1$ and CD36) [319, 320]. VCAM1 is upregulated in hypoxic conditions induced by insufficient oxygen supply to tissues. Increased activity of VCAM1 is injurious to tissues causing severe organ damage. Also, increased platelet activation has been observed which greatly accelerate the process of cell adhesion to the capillaries [321]. Increase in the granulocyte count is considered as a risk factor in elevating the pathology of sickle cell anemia. Sickle cell RBCs and endothelial cells are known to interact with granulocytes which get triggered to produce inflammatory cytokines and cytotoxic molecules causing cellular damage in organs like lungs and brain [322].

Patients suffering from sickle cell anemia experiences severe chest pain, pain in the abdomen, and extremities. Due to the obstruction of blood capillaries, vaso-occlusive crisis (VOC) occurs, which is marked by pain, cell necrosis, ischemia, and organ damage. Other serious complication of sickle cell anemia includes stroke, cognitive impairment, acute lung damage, retinopathy, renal insufficiency, osteonecrosis, and osteomyelitis [323–325]. Factors such as extreme temperature, physical or emotional stress, and contracting infection have been reported to stimulate episodes of pain in sickle cell anemia patients [326]. The quality of life of sickle cell anemia patients is tremendously affected and their life expectancy is remarkably reduced. Blood transfusion and bone marrow transplant are often used for the management of sickle cell anemia and to minimize the risk of its associated clinical complications [327, 328]. HBOT has been recommended for relieving the pathophysiological effects of sickle cell anemia since HbS shows lower affinity for oxygen compared to HbA. Thus, HBOT can be beneficial in acute episodes of anemia by replenishing the availability of oxygen and by preventing ischemia.

7.15.2 HBOT in Management of SCA

HBOT has been tested and proved beneficial in three clinical complications of sickle cell disease, i.e.,vaso-occlusive crisis, ocular surgeries with complications of sickle cell and leg ulcers in patients homozygous for HbS [329–331]. Sickle cell disease induced retinopathy results by the obliteration of retinal arterioles and venules cutting retinal supply and causing retinal detachment. Administration of supplemental oxygen at 70% was found to inhibit retinal cell death in a cat model of retinopathy [332]. Also, oxygen helps to reduce the hypertrophy of Muller cells which is responsible for proliferative vitreo-retinopathy and lowers the complications during surgery [333].

HBOT is helpful in the management of sickle cell anemia by alleviating the symptoms of cellular hypoxia as well as by inducing cytoprotective effects. One of the most important aspects of HBOT induced cellular protection is triggered via nitric oxide synthase (NOS) mediated mechanism. Administration of HBOT results in decreased regional cerebral blood flow. This results in enhanced nitric oxide (NO) production and helps in preventing neuronal excitotoxicity by inhibition of N-methyl D-aspartate (NMDA) receptors [334, 335]. HBOT has also been reported to reduce the harmful effects of cell adhesion and platelet activation by inducing a decrease in cytostatic activity, peroxynitrite synthesis, and inducible NOS (iNOS) mRNA in mouse peritoneal macrophages activated by lipopolysaccharide (LPS) and gamma interferon [336]. HBOT also triggers induction of endothelial NOS (eNOS) which in turn helps in downregulating the adhesion molecule ICAM1 in endothelial cells [337].

Utilizing HBOT in clinical management of sickle cell disease is highly important in acute cases of sickle cell anemia and in cases where blood transfusion is not possible. The recommended protocol for HBOT against sickle cell disease requires administration of 100% oxygen at 2–2.5 ATA pressure for 60–90 min [274, 338]. The clinical symptoms of sickle cell disease can be usually reversed in a single session but HBOT may be repeated according to the clinical presentations or as directed by the physician.

7.16 Decompression Sickness

7.16.1 Pathophysiology and Effects of Decompression Sickness

Decompression sickness (DCS) is a clinical condition that arises due to a sudden decline in ambient pressure. The reduction in atmospheric pressure affects the gases in the body cavity and fluids resulting in the formation of inert gas bubbles due to super-saturation. The bubbles are formed when the decompression is rapid enough to allow tissue inert gas partial pressure to exceed the ambient pressure causing their super-saturation [339]. The trapping of these bubbles in the arteries or veins generate a condition known as arterial gas embolism (AGE) or venous gas embolism (VGE) [340]. There are various clinical manifestations of decompression sickness which affects dysfunction in various organs. DCS can be caused by activities which results in acute decompression. It usually occurs when there is a transition from the atmospheric pressure at ground level in to a lower pressure at a high altitude or decompression experience after returning to ground level after a dive [341]. Similarly, returning to ambient atmospheric pressure after hyperbaric treatment can give rise to DCS.

In the nineteenth century, miners working in the pressurized chambers called "caisson" designed for coal mining complained of joint pain and neurological disturbance after returning to normal pressure conditions. The resulting clinical condition was initially termed as caisson's disease and later recognized as DCS [342]. Major clinical symptoms of DCS include joint pain or bends, neurological deficits, cardio-respiratory problems, and pulmonary edema [343]. The bubbles formed from inert gases due to super-saturation creates serious clinical complications such as occlusion of blood flow, triggering platelet aggregation, complement system activation, activation of inflammatory pathways, endothelial dysfunction, capillary leakage, mechanical disruption of tissues, and initiation of coagulation cascade (Fig. 7.24) [340, 344, 345]. More serious complications of DCS may lead to hemoconcentration and hypotension and manifest as ataxia, motor dysfunction, shock, and death [346]. The patient needs to be assessed for the clinical symptoms of DCS on the basis of time of onset and progression of symptoms, index of gas burden determined as depth-time exposure and any evidence of barotrauma including rapid ascent during diving, holding breath while ascending, dyspnea, and chest pain. The diagnosis and evaluation of DCS is carried out on the basis of an empirical criteria

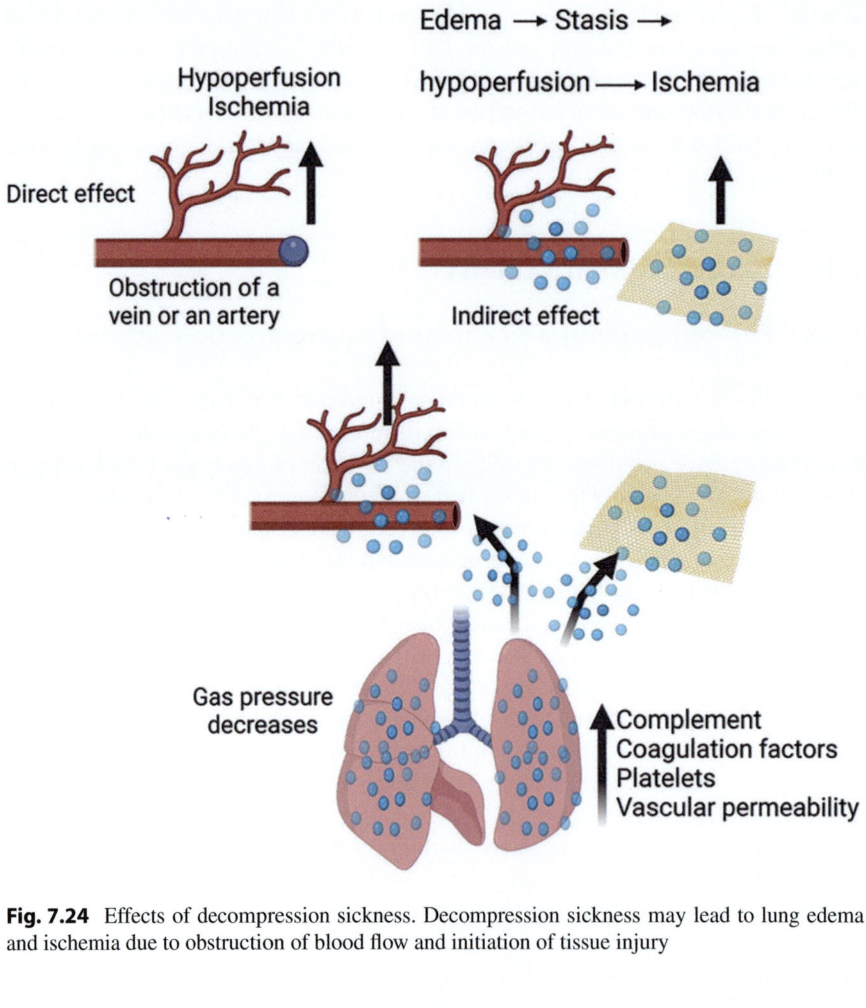

Fig. 7.24 Effects of decompression sickness. Decompression sickness may lead to lung edema and ischemia due to obstruction of blood flow and initiation of tissue injury

involving careful assessment of physiological and neurological symptoms, motor weakness, seizures as well as patient's response upon recompression [347–349]. Patient's clinical history and chest radiography are required to rule out the condition of pneumothorax before starting the recompression. Earlier, recompression with air was used for treatment as it is observed to relieve the patients from the symptoms of DCS. In 1897, a human study using pure oxygen for recompression showed improved results in the form of rapid gas diffusion and bubble resolution [350]. HBOT has been reported to be highly beneficial in the alleviation of DCS related clinical symptoms as it promotes accelerated gas diffusion from bubbles in to tissues along with relieving from tissue ischemia, edema and lessen capillary occlusion by inhibiting neutrophil adhesion to endothelium.

7.16.2 Use of HBOT as a Therapeutic in Treatment of Decompression Sickness

The use of HBOT in DCS treatment is based on the effect of providing 100% oxygen under high pressure to increase the diffusion of inert gas bubble enabling its rapid washout from the body. Hyperbaric oxygen has additional advantages against clinical symptoms of DCS due to the therapeutic nature of oxygen. Most often oxygen is administered in combination of a different inert gas such as helium which is less soluble than the inert gas (mostly nitrogen) in the bubbles. This enhances the rate of bubble resolution without generating neurological deficit in patient which can otherwise manifest by using 100% oxygen at high pressure for prolonged time [351, 352]. Patients showing any signs or symptoms of DCS can be subjected to HBOT. However, supportive measures such as airway protection and monitoring blood pressure are recommended. Also, administration of isotonic IV fluid is recommended during the treatment to overcome the compromised microcirculation and to relieve endothelial damage associated plasma extravasation [353].

Initially, patients are provided with 100% oxygen under normobaric conditions and HBOT is administered after excluding the chances of pneumothorax which need to be managed using tub thoracostomy before starting recompression [346]. The recommended protocol for HBOT in DCS treatment as per US navy Table 5 involves initial recompression at 2.82 ATA with 100% oxygen or oxygen in combination of suitable inert gas for 30 minutes (Fig. 7.25). It is followed by an initial period of 15 minute decompression at 1.9 ATA which is then prolonged for 60 min before the final decompression at normobaric condition (1 ATA) for 15 min. HBOT protocol to manage DCS with neurological manifestations involves 30 minutes of initial decompression at 1.9 ATA instead of 15 min [351]. Majority of cases of DCS shows reversal of clinical symptoms after a single treatment but HBOT sessions may be repeated to ensure the stability of the patient. For treatment of severe symptoms of DCS, 100% oxygen can be administered using a tight fitted mask for 2 h along with aggressive oral rehydration using US navy table 6 [354].

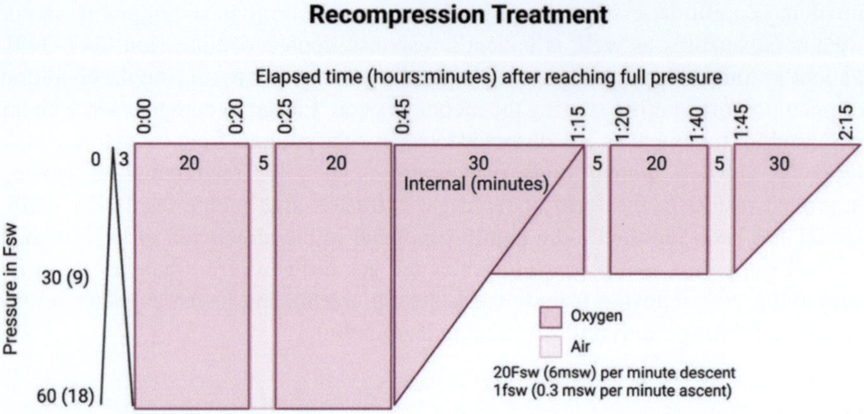

Fig. 7.25 Recompression protocol for decompression sickness

7.17 Intracranial Abscess

7.17.1 Pathogenic Condition and Host Defense in Intracranial Abscess

Intracranial abscess (ICA) is referred to an infection in the cerebral parenchyma which includes pathological conditions such as cerebral abscess, subdural empyema, and epidural empyema. The infection is developed in the form of focal, encapsulated region and can be caused by a variety of microorganisms including bacteria, fungi, protozoa or sometimes helminths [355, 356]. The occurrence of ICA is rare and mostly affects immunocompromised individuals such as patients with HIV or AIDS [357]. Also, patients with malignant condition or undergoing chemotherapy are highly susceptible to fungal infections of brain. Fungal infection such as Aspergillus or members of mucorales family can cause angioinvasive infection leading to septic arteritis and thrombophlebtis [358]. The infection in ICA may originate from hematogenous seeding and cranial trauma. These infections have a tendency to develop in to multiple abscesses. Contiguous infections such as sinusitis, mastoiditis, otitis or dental infections may also be the source of infection in ICA [358, 359]. Development of brain edema and inflammation is associated with ICA due to activation of resident astrocytes and resting microglia in brain (Fig. 7.26). Majority of infections of bacterial origin are caused by *Streptococcus* and *Staphylococcus*. Other anaerobic microorganisms have also been isolated from ICA, which usually originate from normal oral flora. The local environment within the ICA is hypoxic and acidic which promotes the growth and survival of anaerobes. ICA may develop into secondary hypoxic lesions in the surrounding region of brain due to the formation of perifocal edema [360]. Also, the development of abscess in the deeper regions of brain results in inaccessibility of anti-microbial agents. This

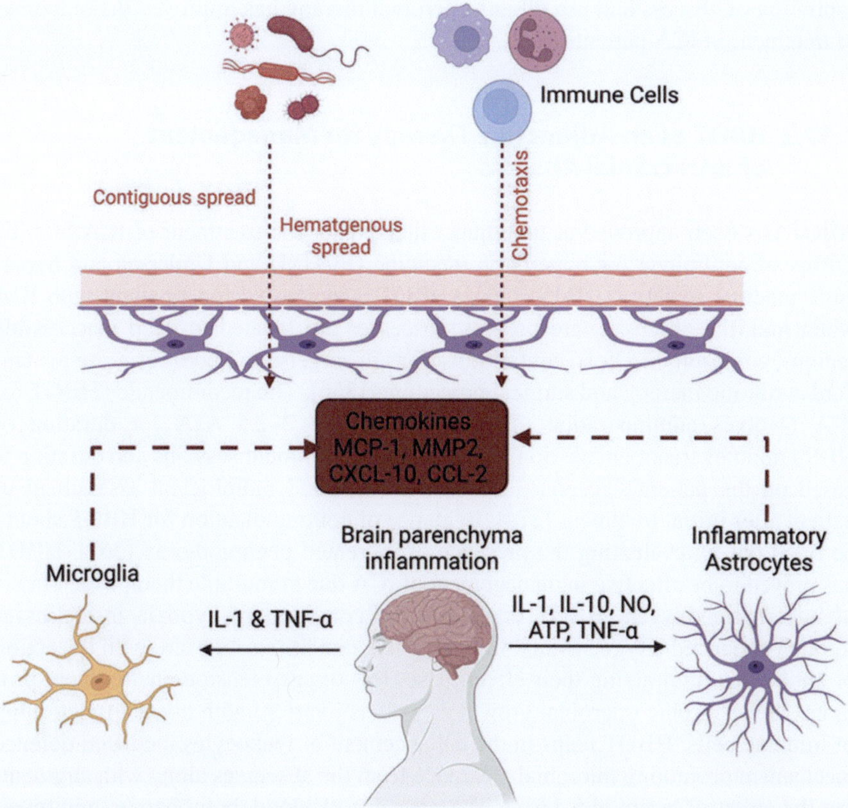

Fig. 7.26 Immunopathology of intracranial abscess. Intracranial abscess may be contracted by mode of contiguous spread or hematogenous spread resulting in the infection of brain parenchyma. Activation of resident astrocytes and resting microglia takes place via recognition of bacterial toll-like receptors 2 (TLR2) resulting in the release of various pro-inflammatory cytokines causing inflammation in the affected region of brain

may have serious life-threatening complications in the form of elevated intracranial pressure and inflammation.

ICA is a rare condition but the mortality rates were typically higher in patients of brain abscess before the advent of better diagnostic, surgical and therapeutic techniques in past few decades [361]. One meta-analysis study showed that the case fatality rate has reduced from 40% to 10% in the past six decades while the patient recovery rate has improved significantly from 33% to 70% [362]. Availability of better imaging techniques like CT scan is considered to be a vital factor in reducing the mortality rate in patients of ICA. According to a retrospective study, the mortality rate in ICA patients has reduced from 40% to 20% in the first decade after the advent of CT scan imaging [363]. Improvement in the imaging techniques has enabled rapid and precise diagnosis of infectious condition. Additionally, introduction of less invasive and more effective surgical procedures like stereotactic aspiration of abscess and use of anti-microbial therapy has improved the outcomes of treatment in ICA patients [364].

7.17.2 HBOT as an Adjunctive Therapy for Management of Intracranial Abscess

HBOT has been approved as an adjunct therapy for the treatment of ICA both by European committee for hyperbaric medicine (ECHM) and Undersea and hyperbaric medical society (UHMS) [365]. HBOT is suggested for a patient with ICA when multiple abscesses are present, abscesses are located in deep inaccessible regions, compromised host, contraindications in surgery, and poor response to standard antibiotic therapy and surgical procedures [366]. The recommended HBOT for ICA involves administration of 100% oxygen at 2–2.5 ATA for duration of 60–90 minutes treatment session. The number of treatment sessions and duration is based on the patient's response to the treatment and radiological assessment of patient after initial treatment [274]. Evidence of contraindication for HBOT should be ruled out by evaluating the presence of untreated pneumothorax [367]. HBOT has proved to be effective in management of ICA due to multiple therapeutic effects of hyperbaric oxygen. HBOT eradicates the condition of hypoxia in abscesses through enhanced oxygen availability. Hypoxic conditions interfere with the action of antibiotics minimizing their effect. Also, low oxygen tension environment promotes the anaerobic microbial growth as well as interfere with normal functioning of immune cells. HBOT helps in the enhancement of leukocytes mediated defence mechanism promoting microbial clearance from the abscesses along with augmenting the effect of antibiotics [368]. The presence of blood–brain barrier minimizes the penetration of antibiotics in the affected region of abscesses. HBOT is known to enable reversible opening of blood–brain barrier, thus promoting increased availability of antibiotics through non-inflamed meninges [369]. Vasoconstriction is mediated under hyperbaric conditions helps in relieving focal brain swelling and reducing of intracranial pressure [370]. It is evident that treatment of ICA has become more plausible after the advent of improved diagnostics tools, effective antibiotics, and utilization of HBOT as an adjunctive measure to manage the complications of disease.

7.18 Femoral Head Necrosis

7.18.1 Pathogenesis and Staging of Femoral Head Necrosis

Femoral head necrosis (FHN) is a clinical condition which arises due to a compromised blood supply to the femoral head (Fig. 7.27). FHN is also known as avascular necrosis, aseptic necrosis or ischemic necrosis signifying the characteristics of the disease [371–373]. Femoral joint is the major weight bearing joint of the body and

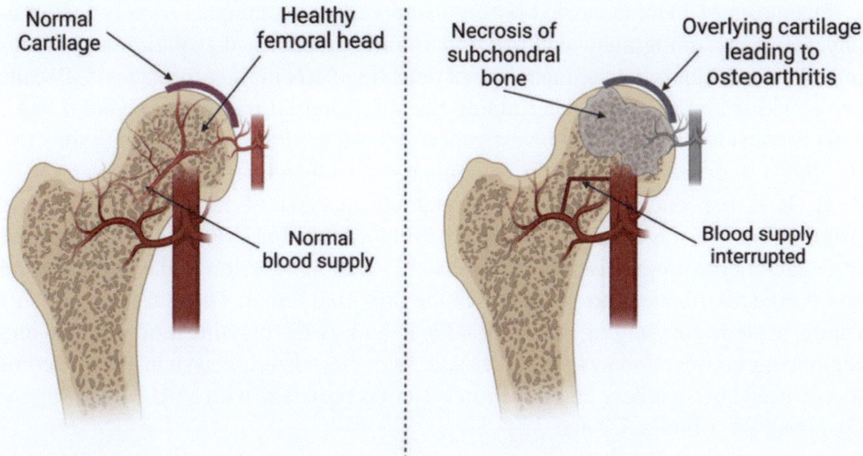

Fig. 7.27 Femoral Head Necrosis. Femoral Head Necrosis is the necrotic condition of femoral head due to disruption in the blood supply to the subchondral region in proximal femur

hence it is highly vascularized for adequate blood supply. There are multiple factors which can hinder the local blood supply to the subchondrial bone of femoral head resulting in osteonecrosis. The structural integrity of the femoral head gets compromised due to insufficient blood supply causing death of osteocytes. Due to the similarity between the ischemia of heart, FHN is sometimes referred as "coronary disease of the hip" [374, 375]. FHN affects relatively young population, usually in the third or fourth decade of their life. As the disease progresses to advanced stages, most of the patients (approx. 70%) require total hip replacement due to destruction of femoral head.

On the basis etiological factors, FHN can be classified as idiopathic or primary FHN (having no clear etiological factor) and secondary FHN (having well identified etiology) [376]. Multiple factors may be responsible for the manifestation of FHN with trauma being the major risk factor. Other etiological factors for FHN includes dietary or environmental factors such as alcohol abuse, cigarette smoking and dysbaric conditions, iatrogenic factors such as corticosteroids, hemodialysis, radiation exposure, cytotoxic treatment or laser surgery and hematologic factors such as hemophilia, sickle cell anemia, polycythemia or thalassemia. Genetic predisposition and metabolic abnormalities leading to osteonecrosis and femoral dysfunction may also lead to the development of FHN [377–379]. The patient experiences limitations in the range of motion and severe pain in groin region which worsens upon bearing weight. Traumatic injuries such as joint dislocation and fracture results in the damage to vascular structure around femoral head hampering the blood supply. Nearly 15–50% cases of femoral neck fractures and 10–25% cases of hip dislocation lead to the development of FHN [380].

Diagnosis of FHN is carried out by using various techniques namely radiography, CT scan, scintigraphy, and MRI. Plain radiographs used as preliminary imaging assessment tools but are incapable of detecting FHN in its early stages. CT scans are sensitive and useful in determining the subchondrial fracture of femoral head. MRI is considered as the best investigation technique for FHN having high sensitivity (99%) and specificity (99%) as compared to other imaging techniques [381–383]. It is especially useful in quantitative analysis of necrotic damage and determining the stage of disease. The staging system for FHN is based on major three factors for prognosis—extent of lesion, location of lesion in the femoral head and presence of bone marrow edema in the proximal femur. There are two classification system for staging of FHN—Ficat and Arlet classification system and Steinberg classification system. Ficat and Arlet classification system is most commonly used but Steinberg classification system is preferred with MRI based prognosis [384, 385] (Tables 7.4 and 7.5).

7.18.2 HBOT as an Aid against Femoral Head Necrosis

There are several medical and surgical interventions for the treatment of FHN aimed to delay the need of total hip replacement. Patients with small lesions are usually subjected to treatment strategies focused on preserving femoral head while for the patients with larger lesions total hip replacement is more suitable option [386]. Biophysical interventions such as use of extracorporeal shock waves and pulsed electromagnetic fields have been used with limited success [387, 388]. Progression of primary femoral osteonecrosis can be prevented by using Enoxaparin while use of drugs like Alendronate for inhibiting the osteoclast activity has been debatable [389, 390]. Surgical procedures for the treatment of FHN focus on either preservation of femoral head or hip arthroplasty. Core decompression is primarily used procedure for preservation of femoral head in the early stage osteonecrosis which helps in reducing intraosseous pressure in the femoral head and restoring vascular flow [391]. In order to provide mechanical support, allograft insertion or non-vascularized autografts are often carried out along with core decompression. Osteotomy can be performed to avoid femoral head collapse which aim at shifting of necrotic regions

Table 7.4 Ficat and Arlet system for staging of FHN

Stage	Symptoms	X-ray findings	Scintigraphy findings
0	None	Normal	Reduced capture
1	None to mild	Normal	Reduced capture
2	Mild	Change in density	Increased capture
2A		Sclerosis or cysts observed	
2B		Applanation or Crescent sign	
3	Mild to Moderate	Loss of sphericity	Increased capture
4	Moderate to Severe	Reduced articular space	Increased capture

Table 7.5 Steinberg classification system for staging of FHN

Stage	Criteria
	Steinberg classification system
0	**Normal or non-diagnostic X-ray, MRI or Bone Scan**
I	**Pain—Normal X-ray—abnormal MRI or Bone Scan**
	A—Mild (<15% of head)
	B—Moderate (15%–30%)
	C—Severe (>30%)
II	**Sclerotic changes or cystic lesions**
	A—Mild (<15%)
	B—Moderate (15%–30%)
	C—Severe (>30%)
III	**Subchondral collapse without flattening**
	A—Mild (<15% of articular surface)
	B—Moderate (15%–30%)
	C—Severe (>30%)
IV	**Flattening of the femoral head**
	A—Mild (<15% of surface and < 2 mm depression)
	B—Moderate (15% - 30% of surface or 2–4 mm depression)
	C—Severe (>30% of surface and > 4 mm depression)
V	**Joint narrowing and/or acetabular involvement**
	A—Mild
	B—Moderate
	C—Severe
VI	**Advanced degenerative changes**

from weight bearing area to a non-weight bearing area of hip joint to transfer the mechanical stress toward healthy bone [392].

HBOT is recommended measure for joint preservation when administered in the early stages of symptomatic FHN. The benefits of HBOT are found to be most prominent in stage 1 and stage 2 of the disease [393, 394]. The characteristic feature of FHN is disruption of blood supply leading to bone tissue hypoxia and ischemia. Administration of HBOT results in the modification of bone metabolism by reducing the dependency of osteogenesis on Hb derived oxygen supply [395]. HBOT enhances the levels of plasma dissolved oxygen, thereby relieving the conditions of hypoxia and ischemia in the affected bone tissues. Additionally, it helps in the reduction of edema in the joints which further facilitates the blood flow and reduction of pain [396]. The recommended HBOT protocol for the treatment of FHN involves administration of 100% oxygen at 2.2–2.5 ATA for 60–120 min. Multiple treatment sessions are required (60–90 sessions) for the improvement in the condition of osteonecrosis [396, 397].

Several studies have been conducted regarding the use of HBOT against FHN. Strauss and Dvorak conducted a meta-analysis study of 15 different research findings or case reports involving a total of 189 patients. It was found that HBOT has a success rate of 97% in 12 months and 81% in 24 months of treatment [398]. Reis et al. performed a review of 12 patients (16 hip joints) characterized by Ficat 1 classification. The patients were given a total of 100 HBOT sessions and followed up for duration of 24 months. It was observed that 13 hip joints were entirely recovered within 8 months of treatment [394]. Another retrospective analysis by Ditri and Montanari on 227 patients of FHN diagnosed with Ficat stage 1 and 2A indicates the effectiveness of HBOT in recovery from the osteonecrotic condition [399]. It is evident from the available reports that treatment of FHN largely depends on the assessment of extent of damage to the femoral head. Early diagnosis and prevention of irreparable damage is critical for the successful treatment of FHN. HBOT has been a useful aid in the preventing the progression of bone necrosis as well as restoring the process of repair and regeneration.

7.19 Thermal Burns

7.19.1 Pathophysiology of Burn Wounds

Thermal burns constitute a complex and dynamic injury resulting in the coagulation of cellular elements in epidermis and dermis. The intensity of heat and duration of heat exposure is critical to the extent of damage which manifests both as local and systemic repercussions [400]. Burn injury is characterized by a central coagulation zone, surrounded by an area of stasis with a border constituting erythema. Heat exposure results in the disruption of vessels and tissues with extravasation of interstitial fluid. There is an increased vasodilation and capillary permeability causing the drainage form microvasculature to the damaged area. Increased inflow of polymorphonuclear neutrophilic leucocytes and monocytes at the site of injury leads to inflammation and edema. As an early event of inflammation, leucocytes and platelets begin to adhere to the vascular endothelial which leads to ischemic conditions in the tissues surrounding injury. Hyperactivation of inflammatory pathways extend their effect to the regions other than the site of injury. Systemic release of vasoactive modulators such as prostaglandins causes edema and inflammation in distant organs [401–403]. There is a high risk of infection due to compromised integrity of skin, impaired immune system and tissue necrosis [404]. The ischemic condition created by occluded capillaries is reinforced by hematological factors such as platelet microthrombi and hemoconcentration. Progressive ischemia sets in during the early days of thermal injury as the surrounding tissues are deprived of oxygen and nutrients compromising their viability [405]. Hypoproteinemia and obliteration of microvascular structure affects the osmotic balance and excessive fluid loss from tissues [406, 407]. Immune system gets downregulated in both cellular and humoral aspects with impaired function of phagocytosis and decreased levels of immunoglobulins respectively [408].

After the initial inflammatory events, formation of new capillaries, collagen fibers, and immature fibroblasts takes place in order to repair the damaged tissues and generate neovasculature. However, the robustness of regenerative mechanisms depends upon attainment of homeostasis first; otherwise, wound healing would be retarded. The patient's condition needs to be stabilized by applying measures to reduce edema, salvaging viable tissues, and replenishing lost fluids. Treatment of burn patient poses a great challenge and the outcome depends on the severity of burn injury, site of injury, and the affected surface. Initial treatment strategies influence the mortality rate of burn patients and usually involve resuscitation of fluid to stabilize patient's situation [409]. Application of topical agents are used to manage pain, edema and reduces the chances of infection. Burn injuries are classified into three different degrees on the basis of severity and depth of cellular destruction (Fig. 7.28) [401, 410]. The burned tissues are carefully excised while the debrided area is covered to avoid fluid loss and infections. Second degree burns or partial thickness burns are associated with massive fluid loss leading to dehydration which can deepen the wound and forming wounds with full thickness loss [411]. Treatment strategies are focused on preserving maximum viable tissues so that natural course of healing can take place. Surgical measures are implemented which involves tangential excision and grafting procedures of second and third degree burns especially in the regions of functional and aesthetic importance.

7.19.2 Role of HBOT in Treatment of Thermal Burns

Effectiveness of HBOT in treatment of burn injuries was first evidence in 1965 when Wada observed that coal miners treated with HBOT for CO poisoning also exhibited rapid healing of second degree burns [412]. Later, in 1970, Gruber et al. determined that administration of HBOT in burn patients restores oxygen levels in hypoxic tissues subadjacent to full thickness injury [413]. HBOT is recommended as a treatment modality in the early stages of second and third degree burns while it is helpful as an adjunctive treatment measure in the later stages of healing. It is suggested that HBOT should be employed within 6 h of injury or earlier to accomplish its maximum benefits [414]. The benefits of HBOT are more apparent in case of burn injuries accompanied by CO poisoning or cyanide poisoning [415]. There are multiple physiological aspects which can be positively altered by the use of HBOT in burn injury treatment. HBOT promotes pre-capillary vasoconstriction which aids in reducing plasma extravasation and reduction of edema. High levels of oxygen at the site of burn injury and surrounding tissues results in relief from ischemic and hypoxic conditions and preservation of viable tissues. Overall microcirculation is improved under hyperoxic conditions while the adhesion of platelets and neutrophils is inhibited. Oxygen has an anti-microbial effect with respect to its ability to promote free radical formation as well as activation of polymorphonuclear leucocytes [416]. HBOT is especially beneficial in the later stages of treatment in

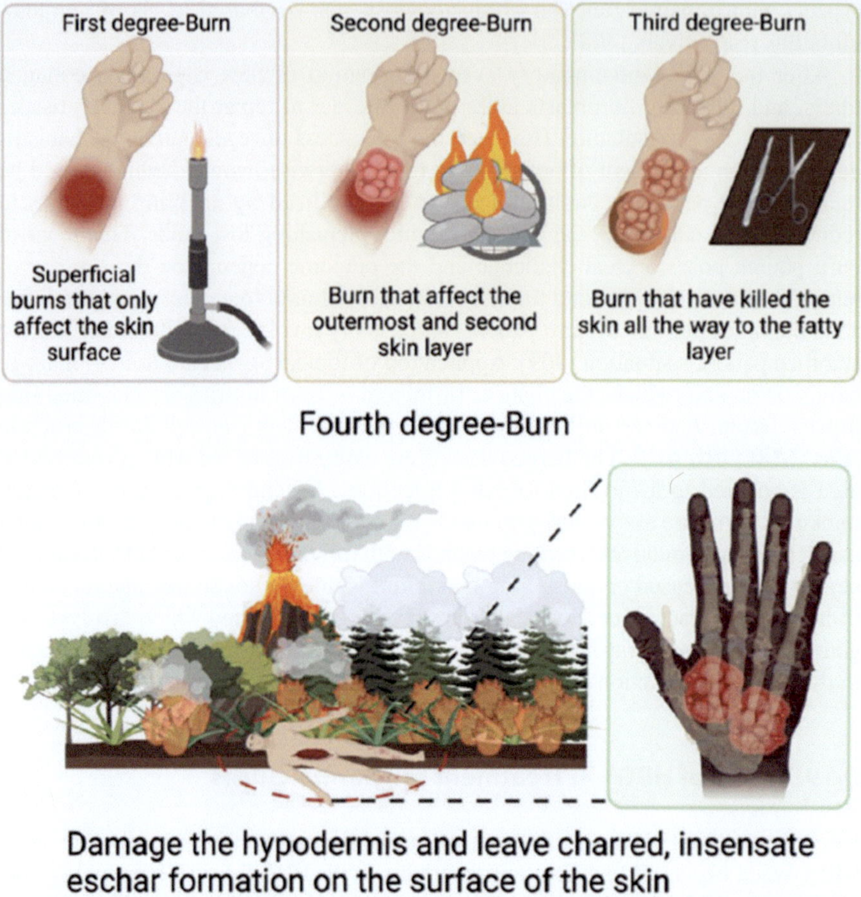

Fig. 7.28 Degrees of thermal burn injuries. Thermal burn injuries are categorized into degrees based on the extent of injury or damage to the affected site of burn injury. First degree burn injuries affects only the outer layer of skin and are relatively mild. Second degree burns affect the outer skin layer and underneath dermis. Blisters will appear on skin which will hurt on touch. Third degree burns, also called as full thickness burn results in complete damage to two outermost skin layers and presents charred appearance. Nerve endings and subcutaneous fat layer are also affected in third degree burn injuries. Fourth degree burns are potentially life-threatening as it affects all the skin layers as well as muscles, bones and tendons

regenerative processes like angiogenesis, collagen formation, and deposition [417]. The effect of HBOT on angiogenesis also increases the chances of graft survival since grafts are dependent on the availability of oxygen and proper blood circulation to ensure their viability. Thus, integration of HBOT with the usual treatment module for burn injuries greatly enhances the positive outcomes of treatment (Fig. 7.29).

HBOT for treatment of burn injuries is recommended to be performed at 2–2.5 ATA of 100% oxygen for up to 2 h. The treatment is usually carried out at a frequency of two sessions daily until the patient's condition is improved [418]. Initial

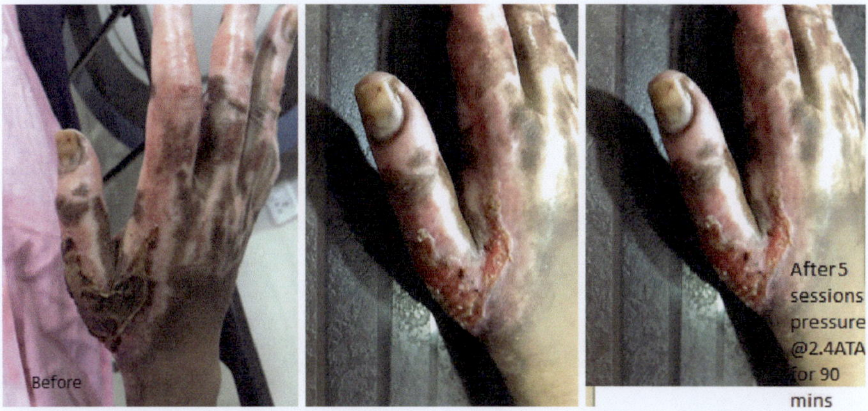

Fig. 7.29 Effect of HBOT on thermal burns

fluid resuscitation is provided to the patient as soon as possible before administration of HBOT and other treatment modalities [418].

7.20 Delayed Radiation Injuries

7.20.1 Pathology and Complications of Delayed Radiation Injuries

Radiation injury has been approved by undersea and hyperbaric medical society as one of the indications where HBOT can be used for treatment [419]. Radiation injuries mainly arise due to the therapeutic use of ionizing radiations for the treatment of cancer patients (Fig.7.30). Cancer patients undergo multitactical treatment plan which includes surgery, chemotherapy and radiation therapy. Both chemotherapy and radiation impart cellular toxicity and acts on malignant cancer cells as well as surrounding healthy tissues.

Ionizing radiations triggers free radical production in cells which eventually suffer DNA damage and undergoes mitotic or reproductive death. As far as the healthy tissues are concerned, radiation induced damage can be observed as acute, subacute or delayed radiation injuries [420]. Acute radiation complications occur in the tissues which have a high rate of cellular loss and mitosis such as oral, gastric, and colorectal mucosa [421, 422]. The severity of acute radiation injury is subjected to the dosages of radiation and treatment time. The patients can be treated symptomatically alongside augmenting their nutritional care. However, in some cases, acute radiation injuries can evolve to develop chronic or delayed radiation injury [420].

Delayed or late radiation injuries are progressive in nature and can limit the clinical dose of radiations. These develop characteristically when the patient experiences only mild acute side effects. There is usually a latent period of 6 months or more and occurrence of any surgery physical trauma or wound triggers the onset of

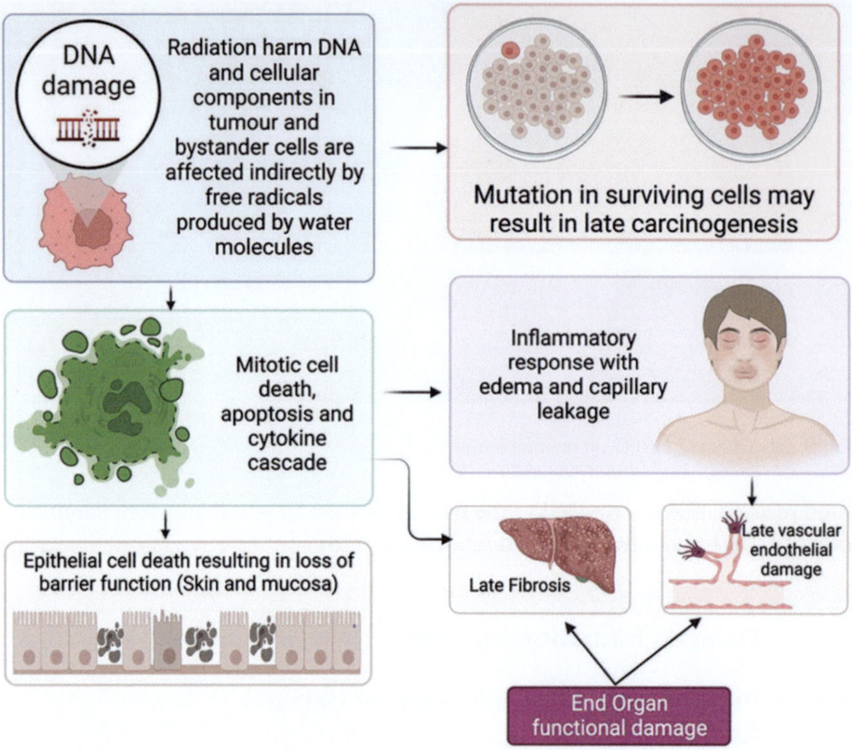

Fig. 7.30 Radiation induced injuries. Radiation therapy results in the increased production of free radicals targeted toward malignant cells. However, healthy cells are also affected in the process causing mitotic cell death apoptosis and necrosis

delay radiation injury [423]. The radiation dosage administered during the course of cancer treatment is based on the balance between tolerance level and minimal dose required for radiation therapy. Complications due to radiation therapy are unavoidable and thus the dosage levels are adjusted such that optimum therapeutic effects are sustained with minimal side effects. The radiation tolerance levels of some patients are lower making them highly susceptible to radiation induced damage. Certain genetic conditions such as ataxia telangiectasia, Nijmegen breakage syndrome, and Fanconi's anemia pose increased risk to patients by lowering their radiation tolerance [424, 425]. Apart from these, collagen vascular disorders and patients subjected to higher dose of radiation than the standard dosage can also increase the chances of radiations induced complications [426]. In order to reduce the extent of damage to healthy tissues during radiation therapy, modern radiation therapy is focused on precise targeting of malignant tissues by using advanced imaging, computing and equipment. The radiation field is modulated continuously as it encounters the targeted area so as to affect only the malignant tissues without causing damage to surrounding normal tissues. Radioprotector drugs like amifostine, which functions as a free radical scavenger, are used in preventing radiation induced injuries [427].

The mechanism of delayed radiation injury was earlier considered to be related to its effect on vascular system characterized by obliterative endarteritis [428]. However, clinical evidences suggest the involvement of bioactive molecules released in response to radiation and implicates damage to normal tissues. These molecules comprise primarily of fibrogenetic cytokines and their effect is termed as fibro-atrophic effect. This model emphasizes that delayed radiation injury is caused due to the depletion of stem cells and parenchymal cells rather than due to the effect of vascular changes. Fleckenstein et al., reported TGF-β as one of the major cytokines associated with radiation injury while interleukins IL-1, IL-2, IL-4, IL5, IL-6, IL-7, IL-8, IL-10, IL-12, IL-13, and IL-17 are some other cytokines involved in radiation induced cytotoxic effect [429].

7.20.2 HBOT as an Aid in Treatment of Delayed Radiation Injuries

HBOT has been widely used in the treatment of delayed radiation injuries and in some conditions known to prevent it when used as a prophylactic intervention [430]. The tissue damage in delayed radiation injury is caused as a result of three different factors—fibrosis, stem cell depletion, and vascular obliteration. HBOT acts on different physiological aspects to counteract the damage induced by these three factors. Hyperbaric and hyperoxic conditions has been demonstrated to enhance overall tissue oxygenation, increasing vascular density, neovascularization, and enhanced mobilization of stem cells to the site of injury to promote regeneration and healing (Fig. 7.31) [431, 432]. Site-specific applications of HBOT in the treatment of delayed radiation injuries have been well studied and documented. The recommended protocol for administration of HBOT in delayed radiation injuries involves

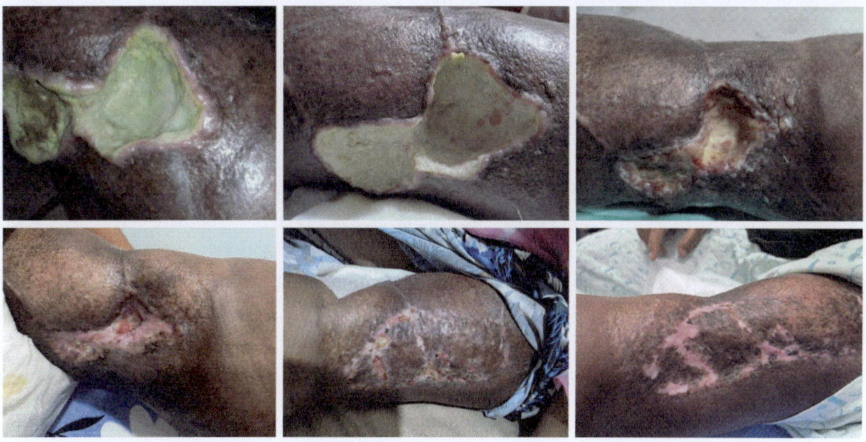

Fig. 7.31 Effect of HBOT on radiation induced wounds

providing the patient with 100% oxygen at 2.4 ATA pressure for 90 min [274]. Multidisciplinary management of delayed radiation injury is required involving surgical intervention, use of radiosensitizer to minimize the radiation damage and administration of HBOT.

7.20.3 Marx's Protocol for Radionecrosis Treatment Using HBOT

At United states Air force (USAF) Medical center, Dr. Robert Marx and colleagues devised a staging system for management of radionecrosis using HBOT in collaboration with Davis and his co-workers from USAF hyperbaric medicine center [433]. The patients were assigned stages on the basis on the extent of radionecrotic damage. Marx's protocol involves administration of HBOT in a multiplace hyperbaric chamber at 2.4 ATA of 100% oxygen for 90 minutes [433, 434]. There are two basic elements which are emphasized in Marx's protocol:

1. Administration of presurgical HBOT (before debridement or resection) having 30 HBOT sessions and an additional 10 HBOT sessions after surgery.
2. Necrotic bone should be surgically extirpated even in case of bone discontinuity or requirement of surgical reconstruction.

The staging system in Marx's protocol for utilization of HBOT in radionecrosis treatment is as follows:

(a) Stage I: Patients with minimal necrotic damage are administered 30 HBOT sessions at 2.4 ATA for 90 min. If the improvement is observed after initial treatment, additional 10 HBOT sessions are provided after debridement.
(b) Stage II: If no improvement is observed in stage I, the patient is advanced to stage II. More formal surgical debridement procedure is required for the patient in order to eliminate all necrotic bone. Patients are administered 30 HBOT sessions at 2.4 ATA for 90 minutes after debridement is carried out. After all necrotic bone is eliminated, an additional 10 HBOT session are administered. If the resection is further required for removing the necrotic bone completely, the patient is advanced to stage III.
(c) Stage III: Patients with orocutaneous fistula, fracture or with necrosis that extends to inferior cortical border of mandible are presented in stage III. The patients are provided 30 HBOT sessions at 2.4 ATA for 90 min prior to resection. Additional 10 HBOT sessions are carried out post-operatively. The patients are prepared for bone graft reconstruction by administering 20 additional HBOT sessions 10 weeks after resection.

References

1. Mechem CC, Manaker S, Traub S. *Hyperbaric oxygen therapy.* UpToDate. https://www.uptodate.com/contents/hyperbaric-oxygen-therapy. Accessed June 2019. 12.
2. Edwards ML. Hyperbaric oxygen therapy. Part 2: application in disease. J Vet Emerg Crit Care. 2010;20(3):289–97.
3. Ismail A. Hyperbaric oxygen therapy as a complementary or alternative therapy for chronic oral and gastrointestinal disorders: a narrative review. Int J Altern and Complement Med. 2020:33–40.
4. Marcinkowska AB, et al. Impact of hyperbaric oxygen therapy on cognitive functions: a systematic review. Neuropsychol Rev. 2022;32(1):99–126.
5. Sahni T, Aggarwal S. Use of hyperbaric oxygen therapy in management of orthopedic disorders. Apollo Med. 2012;9(4):318–22.
6. Thackham JA, McElwain DS, Long RJ. The use of hyperbaric oxygen therapy to treat chronic wounds: a review. Wound Repair Regen. 2008;16(3):321–30.
7. Whalen RE, Saltzman HA. Hyperbaric oxygenation in the treatment of acute myocardial infarction. Prog Cardiovasc Dis. 1968;10(6):575–83.
8. van Hulst RA, Klein J, Lachmann B. Gas embolism: pathophysiology and treatment. Clin Physiol Funct Imaging. 2003;23(5):237–46.
9. Warren B, Philp R, Inwood M. The ultrastructural morphology of air embolism: platelet adhesion to the interface and endothelial damage. Br J Exp Pathol. 1973;54(2):163.
10. Chung T, et al. Platelet activation in acute pulmonary embolism. J Thromb Haemost. 2007;5(5):918–24.
11. Cheng J, et al. Ion channels and vascular diseases. Arterioscler Thromb Vasc Biol. 2019;39(5):e146–56.
12. Moccia F, Berra-Romani R, Tanzi F. Update on vascular endothelial Ca2+ signalling: a tale of ion channels, pumps and transporters. World J Biol Chem. 2012;3(7):127.
13. Hawkins IF. *Potential complications and their prevention*. In: *Carbon dioxide angiography*. CRC Press; 2007. p. 283–298.
14. Leitch D, Green R. Pulmonary barotrauma in divers and the treatment of cerebral arterial gas embolism. Aviat Space Environ Med. 1986;57(10 Pt 1):931–8.
15. Benton P, Francis T, Pethybridge R. Spirometric indices and the risk of pulmonary barotrauma in submarine escape training. Undersea Hyperb Med. 1999;26(4):213.
16. Muth CM, Shank ES. Gas embolism. N Engl J Med. 2000;342(7):476–82.
17. Hare S, et al. Systemic arterial air embolism after percutaneous lung biopsy. Clin Radiol. 2011;66(7):589–96.
18. Voigt P, et al. Systemic air embolism after endoscopy without vessel injury–a summary of reported cases. J Neurol Sci. 2017;376:93–6.
19. Stallworth JM, Martin JB, Postlethwait R. Aspiration of the heart in air embolism. J Am Med Assoc. 1950;143(14):1250–1.
20. O'Quin RJ, Lakshminarayan S. Venous air embolism. Arch Intern Med. 1982;142(12):2173–6.
21. Malik N, et al. Air embolism: diagnosis and management. Futur Cardiol. 2017;13(4):365–78.
22. Davila RM, Rawles T, Mack MJ. Venoarterial air embolus: a complication of vacuum-assisted venous drainage. Ann Thorac Surg. 2001;71(4):1369–71.
23. Souders JE. Pulmonary air embolism. J Clin Monit Comput. 2000;16(5):375–83.
24. van Heerden P, Woods W, Sviri S. Air embolism-a case series and review. Crit Care Resusc. 2004;6:4.
25. Orebaugh SL. Venous air embolism: clinical and experimental considerations. Crit Care Med. 1992;20(8):1169–77.
26. Fukaya E, Hopf HW. HBO and gas embolism. Neurol Res. 2007;29(2):142–5.
27. Calvert JW, Cahill J, Zhang JH. Hyperbaric oxygen and cerebral physiology. Neurol Res. 2007;29(2):132–41.

28. Baiula M, et al. Integrin-mediated adhesive properties of neutrophils are reduced by hyperbaric oxygen therapy in patients with chronic non-healing wound. PLoS One. 2020;15(8):e0237746.
29. Edsell M, Kirk-Bayley J. Hyperbaric oxygen therapy for arterial gas embolism. Br J Anaesth. 2009;103(2):306–7.
30. Benson J, Adkinson C, Collier R. Hyperbaric oxygen therapy of iatrogenic cerebral arterial gas embolism. Undersea Hyperb Med. 2003;30(2):117.
31. Broome J, Smith D. Pneumothorax as a complication of recompression therapy for cerebral arterial gas embolism. Undersea Biomed Res. 1992;19(6):447–55.
32. Tetzloff K, et al. Risk factors for pulmonary barotrauma in divers. Chest. 1997;112(3):654–9.
33. Jain KK, Baydin S. Textbook of hyperbaric medicine. Springer; 2017.
34. Burney RG, et al. Anti-arrhythmic effects of lidocaine metabolites. Am Heart J. 1974;88(6):765–9.
35. Shaikh N, Ummunisa F. Acute management of vascular air embolism. J Emerg Trauma Shock. 2009;2(3):180.
36. Wu L, Wang R. Carbon monoxide: endogenous production, physiological functions, and pharmacological applications. Pharmacol Rev. 2005;57(4):585–630.
37. Guzman JA. Carbon monoxide poisoning. Crit Care Clin. 2012;28(4):537–48.
38. Mehta S, et al. Carbon monoxide poisoning. J Assoc Physicians India. 2001;49:622–5.
39. Hampson NB, Hauff NM. Carboxyhemoglobin levels in carbon monoxide poisoning: do they correlate with the clinical picture? Am J Emerg Med. 2008;26(6):665–9.
40. Icme F, et al. The relationship between blood lactate, carboxy-hemoglobin and clinical status in CO poisoning. Eur Rev Med Pharmacol Sci. 2014;18(3):393–7.
41. Miró Ò, et al. Mitochondrial cytochrome c oxidase inhibition during acute carbon monoxide poisoning. Pharmacol Toxicol. 1998;82(4):199–202.
42. Thom SR, et al. Intravascular neutrophil activation due to carbon monoxide poisoning. Am J Respir Crit Care Med. 2006;174(11):1239–48.
43. Mayr FB, et al. Effects of carbon monoxide inhalation during experimental endotoxemia in humans. Am J Respir Crit Care Med. 2005;171(4):354–60.
44. Cloyd T, Davidson C, LoVecchio F. Carbon monoxide exposure: evaluation and management. Emerg Med Rep. 2021;42:24.
45. Hancock JT, Neill SJ. Nitric oxide: its generation and interactions with other reactive signaling compounds. Plan Theory. 2019;8(2):41.
46. Jang DH, et al. Alterations in mitochondrial respiration and reactive oxygen species in patients poisoned with carbon monoxide treated with hyperbaric oxygen. Intensive Care Med Exp. 2018;6(1):1–14.
47. Guo D, Hu H, Pan S. Oligodendrocyte dysfunction and regeneration failure: a novel hypothesis of delayed encephalopathy after carbon monoxide poisoning. Med Hypotheses. 2020;136:109522.
48. Garg J, et al. Cardiovascular abnormalities in carbon monoxide poisoning. Am J Ther. 2018;25(3):e339–48.
49. Ning K, et al. Neurocognitive sequelae after carbon monoxide poisoning and hyperbaric oxygen therapy. Med Gas Res. 2020;10(1):30.
50. Zengin E, et al. Neuroprotective effects of amantadine for experimental acute carbon monoxide poisoning. Eur Rev Med Pharmacol Sci. 2022;26(19):6919–27.
51. Kim H-H, Choi S. Therapeutic aspects of carbon monoxide in cardiovascular disease. Int J Mol Sci. 2018;19(8):2381.
52. Altintop I, et al. Factors that influence the decision for hyperbaric oxygen therapy (HBOT) in cases of carbon monoxide poisoning: a retrospective study. Ann Burns Fire Disasters. 2018;31(3):168.
53. del Moral-Barbudo B, et al. Current and research therapies for the prevention and treatment of delayed neurological syndrome associated with carbon monoxide poisoning: a narrative review. Hum Exp Toxicol. 2020;39(6):765–72.

54. Eichhorn L, Thudium M, Jüttner B. The diagnosis and treatment of carbon monoxide poisoning. Dtsch Arztebl Int. 2018;115(51–52):863.
55. Thom SR. Options for treatment of carbon monoxide poisoning, including hyperbaric oxygen therapy. Carbon Monoxide. 2019:271–84.
56. Fischer I, Barak B. Molecular and therapeutic aspects of hyperbaric oxygen therapy in neurological conditions. Biomol Ther. 2020;10(9):1247.
57. De Wolde SD, et al. The effects of hyperbaric oxygenation on oxidative stress, inflammation and angiogenesis. Biomol Ther. 2021;11(8):1210.
58. Chang JS, et al. Hyperbaric oxygen exposure attenuates circulating stress biomarkers: a pilot interventional study. Int J Environ Res Public Health. 2020;17(21):7853.
59. Hayreh SS. Central retinal artery occlusion. Ocular Vascular Occlusive Disorders. 2015:239–305.
60. Hayreh SS, Zimmerman MB. Central retinal artery occlusion: visual outcome. Am J Ophthalmol. 2005;140(3):376. e1-376. e
61. Kuwabara T, Cogan DG. Studies of retinal vascular patterns: part I. Normal architecture. Arch Ophthalmol. 1960;64(6):904–11.
62. Spitznas M, Bornfeld N. *The architecture of the most peripheral retinal vessels*, in *Current research in ophthalmic electron microscopy*. Springer; 1977. p. 63–75.
63. Varma D, et al. A review of central retinal artery occlusion: clinical presentation and management. Eye. 2013;27(6):688–97.
64. Rumelt S, Brown GC. Update on treatment of retinal arterial occlusions. Curr Opin Ophthalmol. 2003;14(3):139–41.
65. Werkmeister RM, et al. Retinal oxygen extraction in humans. Sci Rep. 2015;5(1):1–9.
66. Hayreh SS, Kolder HE, Weingeist TA. Central retinal artery occlusion and retinal tolerance time. Ophthalmology. 1980;87(1):75–8.
67. Celebi ARC. Hyperbaric oxygen therapy for central retinal artery occlusion: patient selection and perspectives. Clin Ophthalmol (Auckland, NZ). 2021;15:3443.
68. Biousse V, et al. Thrombolysis for central retinal artery occlusion. J Neuroophthalmol. 2007;27(3):215–30.
69. Hazin R, Dixon JA, Bhatti MT. Thrombolytic therapy in central retinal artery occlusion: cutting edge therapy, standard of care therapy, or impractical therapy? Curr Opin Ophthalmol. 2009;20(3):210–8.
70. Butler F, Chalfin S. The eye in the wilderness. In: Auerbach PS, editor. Wilderness Med. 5th ed. St Louis, MO: Mosby; 2007.
71. Patz A. Oxygen inhalation in retinal arterial occlusion: a preliminary report. Am J Ophthalmol. 1955;40(6):789–95.
72. Yu D-Y, et al. Intraretinal oxygen distribution and consumption during retinal artery occlusion and graded hyperoxic ventilation in the rat. Invest Ophthalmol Vis Sci. 2007;48(5):2290–6.
73. Li HK, Dejean BJ, Tang RA. Reversal of visual loss with hyperbaric oxygen treatment in a patient with Susac syndrome. Ophthalmology. 1996;103(12):2091–8.
74. Duker J, Brown GC. Recovery following acute obstruction of the retinal and choroidal circulations. A case history. Retina (Philadelphia, Pa). 1988;8(4):257–60.
75. Murphy-Lavoie H, Harch P, VanMeter K. Effect of hyperbaric oxygen on central retinal artery occlusion. Australia: UHMS Scientific Assembly; 2004.
76. Hertzog L, et al. Central retinal artery occlusion treated with hyperbaric oxygen. J Hyperbaric Med. 1992;7(1):33–42.
77. Beiran I, et al. Early hyperbaric oxygen therapy for retinal artery occlusion. Eur J Ophthalmol. 2001;11(4):345–50.
78. Rauch SD. Idiopathic sudden sensorineural hearing loss. N Engl J Med. 2008;359(8):833–40.
79. Merchant SN, Adams JC, Nadol JB Jr. Pathology and pathophysiology of idiopathic sudden sensorineural hearing loss. Otol Neurotol. 2005;26(2):151–60.
80. Lazarini PR, Camargo ACK. Idiopathic sudden sensorineural hearing loss: etiopathogenic aspects. Rev Bras Otorrinolaringol. 2006;72:554–61.

81. Marcucci R, et al. Cardiovascular and thrombophilic risk factors for idiopathic sudden sensorineural hearing loss. J Thromb Haemost. 2005;3(5):929–34.
82. Berrocal JG, et al. Role of viral and Mycoplasma pneumoniae infection in idiopathic sudden sensorineural hearing loss. Acta Otolaryngol. 2000;120(7):835–9.
83. Hoch S, Vomhof T, Teymoortash A. Critical evaluation of round window membrane sealing in the treatment of idiopathic sudden unilateral hearing loss. Clin Exp Otorhinolaryngol. 2015;8(1):20–5.
84. Toubi E, et al. Immune-mediated disorders associated with idiopathic sudden sensorineural hearing loss. Ann Otol Rhinol Laryngol. 2004;113(6):445–9.
85. Xie W, et al. Analysis of clinical and laboratory findings of idiopathic sudden sensorineural hearing loss. Sci Rep. 2020;10(1):1–8.
86. Nosrati-Zarenoe R, Hansson M, Hultcrantz E. Assessment of diagnostic approaches to idiopathic sudden sensorineural hearing loss and their influence on treatment and outcome. Acta Otolaryngol. 2010;130(3):384–91.
87. Agarwal L, Pothier DD. Vasodilators and vasoactive substances for idiopathic sudden sensorineural hearing loss. Cochrane Database Syst Rev. 2009;4
88. Awad Z, Huins C, Pothier DD. Antivirals for idiopathic sudden sensorineural hearing loss. Cochrane Database Syst Rev. 2012;8
89. Skarżyńska MB, et al. Effectiveness of various treatments for sudden sensorineural hearing loss—a retrospective study. Life. 2022;12(1):96.
90. Mattox DE, Lyles CA. Idiopathic sudden sensorineural hearing loss. Otol Neurotol. 1989;10(3):242–7.
91. Körpinar Ş, et al. Factors influencing the outcome of idiopathic sudden sensorineural hearing loss treated with hyperbaric oxygen therapy. Eur Arch Otorhinolaryngol. 2011;268(1):41–7.
92. Murphy-Lavoie H, et al. Hyperbaric oxygen therapy for idiopathic sudden sensorineural hearing loss. Undersea Hyperb Med. 2012;39(3):777.
93. Holy R, et al. Hyperbaric oxygen therapy in idiopathic sudden sensorineural hearing loss (ISSNHL) in association with combined treatment. Undersea Hyperb Med. 2011;38(2):137.
94. Bennett MH, et al. Hyperbaric oxygen for idiopathic sudden sensorineural hearing loss and tinnitus. Cochrane Database Syst Rev. 2012;10
95. Chandrasekhar SS, et al. Clinical practice guideline: sudden hearing loss (update). Otolaryngol Head Neck Surg. 2019;161(1_suppl):S1–S45.
96. Olex-Zarychta D. Hyperbaric oxygenation as adjunctive therapy in the treatment of sudden sensorineural hearing loss. Int J Mol Sci. 2020;21(22):8588.
97. Barthelemy A, Rocco M. *Sudden deafness*, in *Handbook on hyperbaric medicine*. Springer; 2006. p. 451–468.
98. Aslan I, et al. Does the addition of hyperbaric oxygen therapy to the conventional treatment modalities influence the outcome of sudden deafness? Otolaryngol Head Neck Surg. 2002;126(2):121–6.
99. Racic G, et al. Hyperbaric oxygen in the treatment of sudden hearing loss. ORL. 2003;65(6):317–20.
100. Fattori B, et al. Alternobaric and hyperbaric oxygen therapy in the immediate and long-term treatment of Meniere's disease. Audiology. 1996;35(6):322–34.
101. Velnar T, Bailey T, Smrkolj V. The wound healing process: an overview of the cellular and molecular mechanisms. J Int Med Res. 2009;37(5):1528–42.
102. Enoch S, Leaper DJ. Basic science of wound healing. Surgery (Oxford). 2008;26(2):31–7.
103. Mathieu D, Linke J-C, Wattel F. *Non-healing wounds*, in *Handbook on hyperbaric medicine*. Springer; 2006. p. 401–428.
104. Medina A, et al. Pathophysiology of chronic nonhealing wounds. J Burn Care Rehabil. 2005;26(4):306–19.
105. Bishop A. Role of oxygen in wound healing. J Wound Care. 2008;17(9):399–402.
106. Schreml S, et al. Oxygen in acute and chronic wound healing. Br J Dermatol. 2010;163(2):257–68.
107. Sen CK. Wound healing essentials: let there be oxygen. Wound Repair Regen. 2009;17(1):1–18.

108. Gilmore M. Phases of wound healing. Dimens Oncol Nurs. 1991;5(3):32–4.
109. Wallace HA, Basehore BM, Zito PM. Wound healing phases. In: StatPearls. StatPearls Publishing, Treasure Island (FL); 2023. Available from https://www.ncbi.nlm.nih.gov/books/NBK470443 PMID: 29262065
110. Rodrigues M, et al. Wound healing: a cellular perspective. Physiol Rev. 2019;99(1):665–706.
111. Yussof SJM, et al. Cellular events and biomarkers of wound healing. Indian J Plast Surg. 2012;45(02):220–8.
112. Park JE, Barbul A. Understanding the role of immune regulation in wound healing. Am J Surg. 2004;187(5):S11–6.
113. Trabold O, et al. Lactate and oxygen constitute a fundamental regulatory mechanism in wound healing. Wound Repair Regen. 2003;11(6):504–9.
114. Hong WX, et al. The role of hypoxia-inducible factor in wound healing. Adv Wound Care. 2014;3(5):390–9.
115. Hoenig MR, Bianchi C, Sellke FW. Hypoxia inducible factor-1α, endothelial progenitor cells, monocytes, cardiovascular risk, wound healing, cobalt and hydralazine: a unifying hypothesis. Curr Drug Targets. 2008;9(5):422–35.
116. Hunt TK, et al. Studies on inflammation and wound healing: angiogenesis and collagen synthesis stimulated in vivo by resident and activated wound macrophages. Surgery. 1984;96(1):48–54.
117. Pastar I, et al. Epithelialization in wound healing: a comprehensive review. Adv Wound Care. 2014;3(7):445–64.
118. Rognoni E, et al. Fibroblast state switching orchestrates dermal maturation and wound healing. Mol Syst Biol. 2018;14(8):e8174.
119. Opalenik SR, Davidson JM. Fibroblast differentiation of bone marrow-derived cells during wound repair. FASEB J. 2005;19(11):1561–3.
120. Weiss JM. Treatment of leg edema and wounds in a patient with severe musculoskeletal injuries. Phys Ther. 1998;78(10):1104–13.
121. Bhutani S, Vishwanath G. Hyperbaric oxygen and wound healing. Indian J Plast Surg. 2012;45(02):316–24.
122. Rodriguez PG, et al. The role of oxygen in wound healing: a review of the literature. Dermatol Surg. 2008;34(9):1159–69.
123. Andrade SMD, Santos ICRV. Hyperbaric oxygen therapy for wound care. Rev Gaucha Enferm. 2016:37.
124. Shukla SK, et al. Pharmacological control of inflammation in wound healing. J Tissue Viability. 2019;28(4):218–22.
125. Lam G, et al. Hyperbaric oxygen therapy: exploring the clinical evidence. Adv Skin Wound Care. 2017;30(4):181–90.
126. Sander AL, et al. In vivo effect of hyperbaric oxygen on wound angiogenesis and epithelialization. Wound Repair Regen. 2009;17(2):179–84.
127. Memar MY, et al. Hyperbaric oxygen therapy: antimicrobial mechanisms and clinical application for infections. Biomed Pharmacother. 2019;109:440–7.
128. Tejada S, et al. Therapeutic effects of hyperbaric oxygen in the process of wound healing. Curr Pharm Des. 2019;25(15):1682–93.
129. Ramasastry SS. Chronic problem wounds. Clin Plast Surg. 1998;25(3):367–96.
130. Guo SA, DiPietro LA. Factors affecting wound healing. J Dent Res. 2010;89(3):219–29.
131. Qing C. The molecular biology in wound healing & non-healing wound. Chin J Traumatol. 2017;20(4):189–93.
132. Uccioli L, et al. Non-healing foot ulcers in diabetic patients: general and local interfering conditions and management options with advanced wound dressings. J Wound Care. 2015;24(Sup4b):35–42.
133. Kalani M, et al. Transcutaneous oxygen tension and toe blood pressure as predictors for outcome of diabetic foot ulcers. Diabetes Care. 1999;22(1):147–51.
134. Trøstrup H, et al. What is new in the understanding of non healing wounds epidemiology, pathophysiology, and therapies. Ulcers. 2013;2013

135. Jeffcoate WJ, et al. Assessing the outcome of the management of diabetic foot ulcers using ulcer-related and person-related measures. Diabetes Care. 2006;29(8):1784–7.
136. Nickinson AT, et al. A systematic review investigating the identification, causes, and outcomes of delays in the management of chronic limb-threatening ischemia and diabetic foot ulceration. J Vasc Surg. 2020;71(2):669–81. e2
137. Mehraj M, Shah I. A review of Wagner classification and current concepts in management of diabetic foot. Int J Orthop Sci. 2018;4(1):933–5.
138. Johnston BR, et al. The mechanism of hyperbaric oxygen therapy in the treatment of chronic wounds and diabetic foot ulcers. RI Med J. 2016;99(2):26–9.
139. Ong M. Hyperbaric oxygen therapy in the management of diabetic lower limb wounds. Singap Med J. 2008;49(2):105.
140. Godman CA, et al. Hyperbaric oxygen induces a cytoprotective and angiogenic response in human microvascular endothelial cells. Cell Stress Chaperones. 2010;15(4):431–42.
141. Gürdöl F et al. *Collagen synthesis, nitric oxide and asymmetric dimethylarginine in diabetic subjects undergoing hyperbaric oxygen therapy.* Physiol Res 59, no. 3. (2010);423–9.
142. Hehenberger K, et al. Dose-dependent hyperbaric oxygen stimulation of human fibroblast proliferation. Wound Repair Regen. 1997;5(2):147–50.
143. Heng MC. Topical hyperbaric therapy for problem skin wounds. J Dermatol Surg Oncol. 1993;19(8):784–93.
144. Gallagher KA, et al. Hyperbaric oxygen and bone marrow–derived endothelial progenitor cells in diabetic wound healing. Vascular. 2006;14(6):328–37.
145. Francis A, Baynosa R. Ischaemia-reperfusion injury and hyperbaric oxygen pathways: a review of cellular mechanisms. Diving Hyperb Med. 2017;47(2):110.
146. Erdoğan A, et al. Efficacy of hyperbaric oxygen therapy in diabetic foot ulcers based on Wagner classification. J Foot Ankle Surg. 2018;57(6):1115–9.
147. Huang ET, et al. A clinical practice guideline for the use of hyperbaric oxygen therapy in the treatment of diabetic foot ulcers. Undersea Hyperb Med. 2015;42(3):205–47.
148. Oliveira N, et al. Treatment of diabetic foot complications with hyperbaric oxygen therapy: a retrospective experience. Foot Ankle Surg. 2014;20(2):140–3.
149. Järbrink K, et al. Prevalence and incidence of chronic wounds and related complications: a protocol for a systematic review. Syst Rev. 2016;5(1):1–6.
150. Bowers S, Franco E. Chronic wounds: evaluation and management. Am Fam Physician. 2020;101(3):159–66.
151. Hsu J-T, et al. Chronic wound assessment and infection detection method. BMC Med Inform Decis Mak. 2019;19(1):1–20.
152. Sen CK, et al. Cutaneous imaging technologies in acute burn and chronic wound care. Plast Reconstr Surg. 2016;138(3 Suppl):119S.
153. Li S, et al. Imaging in chronic wound diagnostics. Adv Wound Care. 2020;9(5):245–63.
154. Percival NJ. Classification of wounds and their management. Surgery (Oxford). 2002;20(5):114–7.
155. Auiwattanakul S, et al. Prevalence of pressure ulcer and nutritional factors affecting wound closure success in Thailand. Mater Sociomed. 2017;29(3):196.
156. Roeckl-Wiedmann I, Bennett M, Kranke P. Systematic review of hyperbaric oxygen in the management of chronic wounds. J Br Surg. 2005;92(1):24–32.
157. Öztürk F, Türel Ermertcan A, İnanır I. Hyperbaric oxygen therapy for the management of chronic wounds. Cutan Ocul Toxicol. 2013;32(1):72–7.
158. Hadanny A, Efrati S. The hyperoxic-hypoxic paradox. Biomol Ther. 2020;10(6):958.
159. Bakker DJ. Clostridial myonecrosis (gas gangrene). Undersea Hyperb Med. 2012;39(3):731.
160. Bryant AE, Stevens DL. Clostridial myonecrosis: new insights in pathogenesis and management. Curr Infect Dis Rep. 2010;12(5):383–91.
161. Stevens DL. The pathogenesis of clostridial myonecrosis. Int J Med Microbiol. 2000;290(4–5):497–502.

162. Darke S, King AM, Slack W. Gas gangrene and related infection: classification, clinical features and aetiology, management and mortality. A report of 88 cases. J Br Surg. 1977;64(2):104–12.
163. Awad MM, et al. Synergistic effects of alpha-toxin and perfringolysin O in Clostridium perfringens-mediated gas gangrene. Infect Immun. 2001;69(12):7904–10.
164. Li J, et al. Clostridium perfringens sporulation and sporulation-associated toxin production. Microbiol Spectr. 2016;4(3):4.3. 08.
165. Navarro MA, McClane BA, Uzal FA. Mechanisms of action and cell death associated with Clostridium perfringens toxins. Toxins. 2018;10(5):212.
166. Buboltz JB, Murphy-Lavoie HM. Gas gangrene. [Updated 2023 Jan 30]. In: StatPearls [Internet]. Treasure Island (FL): StatPearls Publishing; 2023 Jan-. Available from: https://www.ncbi.nlm.nih.gov/books/NBK537030
167. Gupta P, Shah H. *Gas gangrene*. In: *Pediatric musculoskeletal infections*. Springer; 2022. p. 761–771.
168. Sárvári KP, Schoblocher D. The antibiotic susceptibility pattern of gas gangrene-forming clostridium spp. clinical isolates from South-Eastern Hungary. Infect Dis. 2020;52(3):196–201.
169. Stevens DL, et al. Comparison of clindamycin, rifampin, tetracycline, metronidazole, and penicillin for efficacy in prevention of experimental gas gangrene due to Clostridium perfringens. J Infect Dis. 1987;155(2):220–8.
170. Brummelkamp W, Hogendijk J, Boerema I. Treatment of anaerobic infections (clostridial myositis) by drenching the tissues with oxygen under high atmospheric pressure. Surgery. 1961;49(3):299–302.
171. Unsworth IP, Sharp PA. Gas gangrene: an 11-year review of 73 cases managed with hyperbaric oxygen. Med J Aust. 1984;140(5):256–60.
172. Rudge FW. The role of hyperbaric oxygenation in the treatment of clostridial myonecrosis. Mil Med. 1993;158(2):80–3.
173. Yang Z, et al. Interventions for treating gas gangrene. Cochrane Database Syst Rev. 2015;12
174. Korhonen K. *Hyperbaric oxygen therapy in acute necrotizing infections with a special reference to the effects on tissue gas tensions*. in *Annales chirurgiae et gynaecologiae. Supplementum*. 2000.
175. Memar MY, et al. Antimicrobial use of reactive oxygen therapy: current insights. Infect Drug Resist. 2018;11:567.
176. Korhonen K et al. *Management of clostridial gas gangrene and the role of hyperbaric oxygen*. in *Annales chirurgiae et gynaecologiae*. 1999.
177. Stevens DL, Bryant AE. Necrotizing soft-tissue infections. N Engl J Med. 2017;377(23):2253–65.
178. Goldstein EJ, Anaya DA, Dellinger EP. Necrotizing soft-tissue infection: diagnosis and management. Clin Infect Dis. 2007;44(5):705–10.
179. Elliott D, Kufera JA, Myers RA. The microbiology of necrotizing soft tissue infections. Am J Surg. 2000;179(5):361–6.
180. Bulger EM, et al. Impact and progression of organ dysfunction in patients with necrotizing soft tissue infections: a multicenter study. Surg Infect. 2015;16(6):694–701.
181. Bonne SL, Kadri SS. Evaluation and management of necrotizing soft tissue infections. Infect Dis Clin. 2017;31(3):497–511.
182. Kulkarni M, et al. Necrotizing soft-tissue infection: laboratory risk indicator for necrotizing soft tissue infections score. J Lab Physicians. 2014;6(01):046–9.
183. Nichols RL, Florman S. Clinical presentations of soft-tissue infections and surgical site infections. Clin Infect Dis. 2001;33(Supplement_2):S84–93.
184. Lille ST, et al. Necrotizing soft tissue infections: obstacles in diagnosis. J Am Coll Surg. 1996;182(1):7–11.
185. Majeski JA, John JF Jr. Necrotizing soft tissue infections: a guide to early diagnosis and initial therapy. South Med J. 2003;96(9):900–6.
186. Edlich RF, et al. Modern concepts of the diagnosis and treatment of necrotizing fasciitis. J Emerg Med. 2010;39(2):261–5.

187. Moitinho-Silva, L., E. Rodriguez, and S. Weidinger, New perspectives for necrotizing soft-tissue infections pathogen detection. WILEY 111 RIVER ST, HOBOKEN 07030–5774, NJ, USA; 2020, p. 10–10.
188. Puvanendran R, Huey JCM, Pasupathy S. Necrotizing fasciitis. Can Fam Physician. 2009;55(10):981–7.
189. Bouvet A. *Cellulitis and necrotizing fasciitis: microbiology and pathogenesis*. in *Annales de Dermatologie et de Venereologie*. 2001.
190. Meleney FL. Hemolytic streptococcus gangrene. Arch Surg. 1924;9(2):317–64.
191. Ahn C. Necrotizing fasciitis: reviewing the causes and treatment strategies. Adv Skin Wound Care. 2007;20(5):288–93.
192. Brook I, Frazier EH. Clinical features and aerobic and anaerobic microbiological characteristics of cellulitis. Arch Surg. 1995;130(7):786–92.
193. Qvist G. Anaerobic cellulitis and gas gangrene. Br Med J. 1941;2(4206):217.
194. Phan HH, Cocanour CS. Necrotizing soft tissue infections in the intensive care unit. Crit Care Med. 2010;38(9):S460–8.
195. Kobayashi L, et al. Necrotizing soft tissue infections: delayed surgical treatment is associated with increased number of surgical debridements and morbidity. J Trauma Acute Care Surg. 2011;71(5):1400–5.
196. Aragón-Sánchez J, et al. Necrotizing soft-tissue infections in the feet of patients with diabetes: outcome of surgical treatment and factors associated with limb loss and mortality. Int J Low Extrem Wounds. 2009;8(3):141–6.
197. Mortensen CR. Hyperbaric oxygen therapy. Curr Anaesth Crit Care. 2008;19(5–6):333–7.
198. Massey PR, et al. Hyperbaric oxygen therapy in necrotizing soft tissue infections. J Surg Res. 2012;177(1):146–51.
199. Howell GM, Rosengart MR. Necrotizing soft tissue infections. Surg Infect. 2011;12(3):185–90.
200. Soh CR, et al. Hyperbaric oxygen therapy in necrotising soft tissue infections: a study of patients in the United States Nationwide inpatient sample. Intensive Care Med. 2012;38(7):1143–51.
201. Hedetoft M, Bennett MH, Hyldegaard O. Adjunctive hyperbaric oxygen treatment for necrotising soft-tissue infections: a systematic review and meta-analysis. Diving Hyperb Med. 2021;51(1):34.
202. Schmitt SK. Osteomyelitis. Infect Dis Clin. 2017;31(2):325–38.
203. Lew DP, Waldvogel FA. Osteomyelitis. Lancet. 2004;364(9431):369–79.
204. Hart BB. Refractory osteomyelitis. The Hyperbaric oxygen therapy committee report, 2003: p. 79–85.
205. Chung K-P, et al. Refractory bacteremia and osteomyelitis resulting in fatal bacteremic pneumonia with multiorgan failure caused by Mycobacterium simiae in a non-human immunodeficiency virus-infected adult. J Clin Microbiol. 2009;47(11):3777–9.
206. Delasotta LA, et al. Hyperbaric oxygen for osteomyelitis in a compromised host. Open Orthop J. 2013;7:114.
207. Kaplan SL. Osteomyelitis in children. Infect Dis Clin. 2005;19(4):787–97.
208. Hanley ME, Hendriksen S, Cooper JS. *Hyperbaric treatment of chronic refractory osteomyelitis*, in *StatPearls [Internet]*. StatPearls Publishing; 2022.
209. Esterhai JL Jr, et al. Adjunctive hyperbaric oxygen therapy in the treatment of chronic refractory osteomyelitis. J Trauma Acute Care Surg. 1987;27(7):763–8.
210. Chen C-E, et al. Hyperbaric oxygen therapy in the treatment of chronic refractory osteomyelitis: a preliminary report. Chang Gung Med J. 2003;26(2):114–21.
211. Jiang X, et al. Limb salvage and prevention of ulcer recurrence in a chronic refractory diabetic foot osteomyelitis. Diab Metab Syndr Obes Targets Therapy. 2020;13:2289.
212. Mader JT, Shirtliff M, Calhoun JH. Staging and staging application in osteomyelitis. Clin Infect Dis. 1997;25(6):1303–9.
213. Zimmerli W, Flückiger U. Classification and microbiology of osteomyelitis. Der Orthopade. 2004;33(3):267–72.

214. Cierny G III. Surgical treatment of osteomyelitis. Plast Reconstr Surg. 2011;127:190S–204S.
215. Ciampolini J, Harding K. Pathophysiology of chronic bacterial osteomyelitis. Why do antibiotics fail so often? Postgrad Med J. 2000;76(898):479–83.
216. Roy M, et al. Pathophysiology and pathogenesis of osteomyelitis. INTECH Open Access Publisher; 2012. p. 1–26.
217. Zalavras CG, Singh A, Patzakis MJ. Novel technique for medullary canal debridement in tibia and femur osteomyelitis. Clin Orthop Relat Res. 2007;461:31–4.
218. Kim Y-K, et al. Chronic refractory osteomyelitis of the mandible and maxilla related to dental implant placement. Asian J Oral Maxillofac Surg. 2008;20(4):189–92.
219. Yu W, et al. The endothelial-mesenchymal transition (EndMT) and tissue regeneration. Curr Stem Cell Res Ther. 2014;9(3):196–204.
220. Sunkari VG, et al. Hyperbaric oxygen therapy activates hypoxia-inducible factor 1 (HIF-1), which contributes to improved wound healing in diabetic mice. Wound Repair Regen. 2015;23(1):98–103.
221. Rossi F, Della Bianca V, de Togni P. Mechanisms and functions of the oxygen radicals producing respiration of phagocytes. Comp Immunol Microbiol Infect Dis. 1985;8(2):187–204.
222. Fantone JC, Ward P. Role of oxygen-derived free radicals and metabolites in leukocyte-dependent inflammatory reactions. Am J Pathol. 1982;107(3):395.
223. Löndahl M. Hyperbaric oxygen therapy as adjunctive treatment for diabetic foot ulcers. Int J Low Extrem Wounds. 2013;12(2):152–7.
224. Al-Hadi, Hadil Wael. "The effect of hyperbaric oxygen therapy on osteoclast and osteoblast function." PhD diss., Plymouth University, 2013.
225. Carfrae MJ, Kesser BW. Malignant otitis externa. Otolaryngol Clin N Am. 2008;41(3):537–49.
226. Hariga I, et al. Necrotizing otitis externa: 19 cases' report. Eur Arch Otorhinolaryngol. 2010;267(8):1193–8.
227. Clark JH, et al. Malignant otitis externa caused by aspergillus fumigatus: a case report. Otol Neurotol. 2011;32(3):e22–3.
228. Omran AA, El Garem HF, Al Alem RK. Recurrent malignant otitis externa: management and outcome. Eur Arch Otorhinolaryngol. 2012;269(3):807–11.
229. Singh J, Bhardwaj B. The role of surgical debridement in cases of refractory malignant otitis externa. Indian J Otolaryngol Head Neck Surg. 2018;70(4):549–54.
230. Holten KB, Gick J. Management of the patient with otitis externa. J Fam Pract. 2001;50(4):353.
231. Kumar SP, et al. Malignant otitis externa: an emerging scourge. J Clin Gerontol Geriatr. 2013;4(4):128–31.
232. Bhandary S, Karki P, Sinha BK. Malignant otitis externa: a review. Pac Health Dialog. 2002;9(1):64–7.
233. Van Kroonenburgh A, et al. Advanced imaging techniques in skull base osteomyelitis due to malignant otitis externa. Curr Radiol Rep. 2018;6(1):1–14.
234. Magliocca KR, Vivas EX, Griffith CC. Idiopathic, infectious and reactive lesions of the ear and temporal bone. Head Neck Pathol. 2018;12(3):328–49.
235. Karaman E, et al. Malignant otitis externa. J Craniofac Surg. 2012;23(6):1748–51.
236. Cole L, Nuttall T. Clinical techniques: when and how to do a myringotomy–a practical guide. Vet Dermatol. 2021;32(3):302–e82.
237. Finitzo T, et al. Tympanometry and otoscopy prior to myringotomy: issues in diagnosis of otitis media. Int J Pediatr Otorhinolaryngol. 1992;24(2):101–10.
238. Rovers MM, et al. Otitis media. Lancet. 2004;363(9407):465–73.
239. Siyabi AA, et al. Management of malignant Otitis Externa with hyperbaric oxygen therapy: a case series of 20 patients. Inflammation. 2023;2:12.
240. Bath A, Rowe J, Innes A. Malignant otitis externa with optic neuritis. J Laryngol Otol. 1998;112(3):274–7.
241. Gilain L, et al. *Value of hyperbaric oxygen therapy in the treatment of malignant otitis externa. Apropos of a case*. in *Annales D'oto-laryngologie et de Chirurgie Cervico Faciale: Bulletin de la Societe D'oto-laryngologie des Hopitaux de Paris*. 1993.

242. Shupak A, et al. Hyperbaric oxygenation for necrotizing (malignant) otitis externa. Arch Otolaryngol Head Neck Surg. 1989;115(12):1470–5.
243. Davis JC, et al. Adjuvant hyperbaric oxygen in malignant external otitis. Arch Otolaryngol Head Neck Surg. 1992;118(1):89–93.
244. Martel J, et al. *Malignant or necrotizing otitis externa: experience in 22 cases*. in *Annales D'oto-laryngologie et de Chirurgie Cervico Faciale: Bulletin de la Societe D'oto-laryngologie des Hopitaux de Paris*. 2000.
245. Tisch M, et al. The treatment of necrotizing otitis externa with a combination of surgery, antibiotics, specific immunoglobulins and hyperbaric oxygen therapy. HNO. 2003;51(4):315–20.
246. Phillips JS, Jones SE. Hyperbaric oxygen as an adjuvant treatment for malignant otitis externa. Cochrane Database Syst Rev. 2013;5
247. Mahdyoun P, et al. Necrotizing otitis externa: a systematic review. Otol Neurotol. 2013;34(4):620–9.
248. Petruzzelli GJ, Johnson JT. Skin grafts. Otolaryngol Clin N Am. 1994;27(1):25–37.
249. Wei F-C, Mardini S. Flaps and reconstructive surgery. Elsevier Health Sciences; 2016.
250. Hallock GG, Morris SF. Skin grafts and local flaps. Plast Reconstr Surg. 2011;127(1):5e–22e.
251. Andreassi A, et al. Classification and pathophysiology of skin grafts. Clin Dermatol. 2005;23(4):332–7.
252. Zhang F, Lineaweaver W. Acute and sustained effects of vascular endothelial growth factor on survival of flaps and skin grafts. Ann Plast Surg. 2011;66(5):581–2.
253. Robson MC, Krizek TJ. Predicting skin graft survival. Plast Reconstr Surg. 1973;52(3):330.
254. Kakinoki R, et al. Factors affecting the success of arterialised venous flaps in the hand. Injury. 2008;39:18–24.
255. Fukui A, et al. Venous flap—its classification and clinical applications. Microsurgery. 1994;15(8):571–8.
256. Patel KG, Sykes JM. Concepts in local flap design and classification. Oper Tech Otolaryngol Head Neck Surg. 2011;22(1):13–23.
257. Shibuya N, Jupiter DC. Bone graft substitute: allograft and xenograft. Clin Podiatr Med Surg. 2015;32(1):21–34.
258. Ciresi KF, Mathes SJ. The classification of flaps. Orthop Clin N Am. 1993;24(3):383–91.
259. Hallock GG, Ahmadzadeh R, Morris S. Classification of flaps. Flaps Reconstr Surg. 2009:1.
260. Hallock GG. The complete classification of flaps. Microsurgery. 2004;24(3):157–61.
261. Zamboni WA, Baynosa RC. Compromised grafts and flaps. Physiol Med Hyperbaric Oxygen Therapy. 2008:373–95.
262. Cuomo R, et al. Surgical implications of ischemia reperfusion damage and future perspectives. J Investig Surg. 2021;34(1):118–9.
263. Cheng Y-K, et al. Factors affecting graft infection after cranioplasty. J Clin Neurosci. 2008;15(10):1115–9.
264. Fenlon MR, et al. Factors affecting survival and usefulness of implants placed in vascularized free composite grafts used in post–head and neck cancer reconstruction. Clin Implant Dent Relat Res. 2012;14(2):266–72.
265. Stepnick DW, Hayden RE. Postoperative monitoring and salvage of microvascular free flaps. Otolaryngol Clin N Am. 1994;27(6):1201–17.
266. Capla JM, et al. Skin graft vascularization involves precisely regulated regression and replacement of endothelial cells through both angiogenesis and vasculogenesis. Plast Reconstr Surg. 2006;117(3):836–44.
267. Francis A, Baynosa RC. Hyperbaric oxygen therapy for the compromised graft or flap. Adv Wound Care. 2017;6(1):23–32.
268. Finseth F. Clinical salvage of three failing skin flaps by treatment with a vasodilator drug. Plast Reconstr Surg. 1979;63(3):304–8.
269. Yii NW, et al. Thrombolytic therapy:: what is its role in free flap salvage? Ann Plast Surg. 2001;46(6):601–4.
270. Aden KK, Biel MA. The evaluation of pharmacologic agents on composite graft survival. Arch Otolaryngol Head Neck Surg. 1992;118(2):175–8.

271. Maeda M, Fukui A, Tamai S. Combined therapy with antithrombotic agents and radical scavengers for reperfusion injury of flaps. J Reconstr Microsurg. 1991;7(03):233–43.
272. Baynosa RC, Zamboni WA. The effect of hyperbaric oxygen on compromised grafts and flaps. Undersea Hyperb Med. 2012;39(4):857.
273. Wright J. Hyperbaric oxygen therapy for wound healing. World Wide Wounds; 2001.
274. Neuman TS, Thom SR. Physiology and medicine of hyperbaric oxygen therapy. Elsevier Health Sciences; 2008.
275. Phillips JC. Understanding hyperbaric oxygen therapy and its use in the treatment of compromised skin grafts and flaps. Plast Aesthet Nurs. 2005;25(2):72–80.
276. Bada AM, Pope GH. Use of hyperbaric oxygen as adjunct in salvage of near-complete ear amputation. Plast Reconstr Surg Glob Open. 2013;1(1):e1.
277. Renner G, et al. Enhancement of auricular composite graft survival with hyperbaric oxygen therapy. Arch Facial Plast Surg. 2002;4(2):102–4.
278. Rajagopalan S. Crush injuries and the crush syndrome. Med J Armed Forces India. 2010;66(4):317–20.
279. Sahjian M, Frakes M. Crush injuries: pathophysiology and current treatment. Adv Emerg Nurs J. 2007;29(2):145–50.
280. Smith J, Greaves I. Crush injury and crush syndrome: a review. J Trauma Acute Care Surg. 2003;54(5):S226–30.
281. Ziv I, et al. Immediate care of crush injuries and compartment syndromes with the split-thickness skin excision. Clin Orthop Relat Res. 1990;256:224–8.
282. Kemmer A. *Crush injury and other acute traumatic ischemia*, In: *Handbook on Hyperbaric Medicine*. Springer; 2006 p. 305–327.
283. Goodman AD, Got CJ, Weiss A-PC. Crush injuries of the hand. J Hand Surg Am. 2017;42(6):456–63.
284. Haines LN, Doucet JJ. Severe crush injury in adults. 2019. Available from https://www.uptodate.com/contents/severe-crush-injury-in-adults (Accessed date: 30.01. 2022).
285. Agrawal A. Unified classification of open fractures: based on Gustilo and OTA classification schemes. Injury. 2018;49(8):1526–31.
286. Kim PH, Leopold SS. Gustilo-Anderson classification. Springer; 2012.
287. Better OS, Rubinstein I, Reis DN. Muscle crush compartment syndrome: fulminant local edema with threatening systemic effects. Kidney Int. 2003;63(3):1155–7.
288. Myers RA. Hyperbaric oxygen therapy for trauma: crush injury, compartment syndrome, and other acute traumatic peripheral ischemias. Int Anesthesiol Clin. 2000;38(1):139–51.
289. Lv Q, et al. The role of Alpha-1-acid glycoprotein in the diagnosis and treatment of crush syndrome-induced acute kidney injury. Shock. 2021;56(6):1028–39.
290. Durkin A, et al. Contemporary management of pelvic fractures. Am J Surg. 2006;192(2):211–23.
291. Heetveld MJ, et al. Guidelines for the management of haemodynamically unstable pelvic fracture patients. ANZ J Surg. 2004;74(7):520–9.
292. Del Piñal F. An update on the management of severe crush injury to the forearm and hand. Clin Plast Surg. 2020;47(4):461–89.
293. Anderson SR, Vincent JA, Wimalawansa SM. Multi-staged management of a mangled, exploded hand crush injury with long-term outcome: the critical link between surgery and therapy. BMJ Case Rep. 2022;15(5):e248429.
294. Buettner MF, Wolkenhauer D. Hyperbaric oxygen therapy in the treatment of open fractures and crush injuries. Emerg Med Clin North Am. 2007;25(1):177–88.
295. Goldman RJ. Hyperbaric oxygen therapy for wound healing and limb salvage: a systematic review. PM&R. 2009;1(5):471–89.
296. Dougherty JE. The role of hyperbaric oxygen therapy in crush injuries. Crit Care Nurs Q. 2013;36(3):299–309.
297. Sheffield P. Physiological and pharmacological basis of hyperbaric oxygen therapy. Hyperbaric Surgery. 2002:63–109.

298. Hunt T, Mp P. *The effect of varying ambient oxygen tensions on wound metabolism and collagen synthesis;* Surg Gynecol Obstet. 1972;135:561–7.
299. Lopez A, et al. Iron deficiency anaemia. Lancet. 2016;387(10021):907–16.
300. Campbell K. Anaemia: causes and treatment. Nurs Times. 2003;99(43):30–3.
301. Hayden RA. CE credit: what keeps oxygenation on track? Am J Nurs. 1992;92(12):32–42.
302. Billett HH. *Hemoglobin and hematocrit.* Clinical Methods: The History, Physical, and Laboratory Examinations. 3rd edition; 1990.
303. Greenburg AG. Pathophysiology of anemia. Am J Med. 1996;101(2):7S–11S.
304. Ludwig H, Strasser K. *Symptomatology of anemia.* In: *Seminars in oncology*. Elsevier; 2001.
305. English M, et al. Lactic acidosis and oxygen debt in African children with severe anaemia. QJM. 1997;90(9):563–9.
306. Grant MJC, Huether SE, Witte MK. Effect of red blood cell transfusion on oxygen consumption in the anemic pediatric patient. Pediatr Crit Care Med. 2003;4(4):459–64.
307. Boerema I. Life without blood. J Cardiovasc Surg. 1960;1:133–46.
308. Van Meter KW. The effect of hyperbaric oxygen on severe anemia. Undersea Hyperb Med. 2012;39(5):937.
309. Myking O, Schreiner A. Hyperbaric oxygen in hemolytic crisis. JAMA. 1974;227(10):1161–2.
310. Lonergan GJ, Cline DB, Abbondanzo SL. Sickle cell anemia. Radiographics. 2001;21(4):971–94.
311. Platt OS. Sickle cell anemia as an inflammatory disease. J Clin Invest. 2000;106(3):337–8.
312. Rucknagel DL. The genetics of sickle cell anemia and related syndromes. Arch Intern Med. 1974;133(4):595–606.
313. Higgs DR, Wood WG. Genetic complexity in sickle cell disease. Proc Natl Acad Sci. 2008;105(33):11595–6.
314. Vekilov PG. Sickle-cell haemoglobin polymerization: is it the primary pathogenic event of sickle-cell anaemia? Br J Haematol. 2007;139(2):173–84.
315. Maciaszek JL, Andemariam B, Lykotrafitis G. Microelasticity of red blood cells in sickle cell disease. J Strain Anal Eng Des. 2011;46(5):368–79.
316. Harris JW, et al. Studies on the destruction of red blood cells: X. The biophysics and biology of sickle-cell disease. AMA Arch Intern Med. 1956;97(2):145–68.
317. Nader E, Romana M, Connes P. The red blood cell—inflammation vicious circle in sickle cell disease. Front Immunol. 2020;11:454.
318. Lipowsky HH, Sheikh NU, Katz D. Intravital microscopy of capillary hemodynamics in sickle cell disease. J Clin Invest. 1987;80(1):117–27.
319. Cartron J-P, Elion J. Erythroid adhesion molecules in sickle cell disease: effect of hydroxyurea. Transfus Clin Biol. 2008;15(1–2):39–50.
320. Setty B, Stuart MJ. *Vascular cell adhesion molecule-1 is involved in mediating hypoxia-induced sickle red blood cell adherence to endothelium: potential role in sickle cell disease.* 1996: 2311–20.
321. Tomer A, et al. Thrombogenesis in sickle cell disease. J Lab Clin Med. 2001;137(6):398–407.
322. Conran N, et al. Leukocyte numbers correlate with plasma levels of granulocyte–macrophage colony-stimulating factor in sickle cell disease. Ann Hematol. 2007;86(4):255–61.
323. Rees DC, Williams TN, Gladwin MT. Sickle-cell disease. Lancet. 2010;376(9757):2018–31.
324. Kato GJ, et al. Sickle cell disease. Nat Rev Dis Primers. 2018;4(1):1–22.
325. Anie KA. Psychological complications in sickle cell disease. Br J Haematol. 2005;129(6):723–9.
326. Lee L, et al. Adults with sickle cell disease: an interdisciplinary approach to home care and self-care management with a case study. Home Healthc Nurse. 2012;30(3):172–83.
327. Vermylen C, Cornu G. Bone marrow transplantation for sickle cell disease. J Pediatr Hematol Oncol. 1994;16(1):18–21.
328. Drasar E, et al. Blood transfusion usage among adults with sickle cell disease–a single institution experience over ten years. Br J Haematol. 2011;152(6):766–70.

329. Canan H, Ulas B, Altan-Yaycioglu R. Hyperbaric oxygen therapy in combination with systemic treatment of sickle cell disease presenting as central retinal artery occlusion: a case report. J Med Case Rep. 2014;8(1):1–3.
330. Saber AA, et al. A new approach in the management of chronic nonhealing leg ulcers. J Investig Surg. 2005;18(6):321–3.
331. Oguz H, Sobaci G. The use of hyperbaric oxygen therapy in ophthalmology. Surv Ophthalmol. 2008;53(2):112–20.
332. Mervin K, et al. Limiting photoreceptor death and deconstruction during experimental retinal detachment: the value of oxygen supplementation. Am J Ophthalmol. 1999;128(2):155–64.
333. Lewis G, et al. Limiting the proliferation and reactivity of retinal Müller cells during experimental retinal detachment: the value of oxygen supplementation. Am J Ophthalmol. 1999;128(2):165–72.
334. Förstermann U, Sessa WC. Nitric oxide synthases: regulation and function. Eur Heart J. 2012;33(7):829–37.
335. Baynosa RC, et al. The effect of hyperbaric oxygen on nitric oxide synthase activity and expression in ischemia-reperfusion injury. J Surg Res. 2013;183(1):355–61.
336. Louis CA, et al. Distinct arginase isoforms expressed in primary and transformed macrophages: regulation by oxygen tension. Am J Phys Regul Integr Comp Phys. 1998;274(3):R775–82.
337. Buras JA, et al. Hyperbaric oxygen downregulates ICAM-1 expression induced by hypoxia and hypoglycemia: the role of NOS. Am J Phys Cell Phys. 2000;278:C292-302.
338. Reynolds JD. Painful sickle cell crisis: successful treatment with hyperbaric oxygen therapy. JAMA. 1971;216(12):1977–8.
339. Mahon RT, Regis DP. Decompression and decompression sickness. Compr Physiol. 2011;4(3):1157–75.
340. Neuman TS. Arterial gas embolism and decompression sickness. Physiology. 2002;17(2):77–81.
341. Rivera JC. *Decompression sickness among divers: an analysis of 935 cases*. 1963, NAVY EXPERIMENTAL DIVING UNIT PANAMA CITY FL.
342. Leach J et al. *30 Memory and personality of caisson workers with a history of decompression illness*. in *Engineering and Health in Compressed Air Work: Proceedings of the International Conference, Oxford, September 1992*. Taylor & Francis 1994.
343. Vann RD, et al. Decompression illness. Lancet. 2011;377(9760):153–64.
344. Boussuges A, et al. Activation of coagulation in decompression illness. Aviat Space Environ Med. 1998;69(2):129–32.
345. Mitchell SJ, Bennett MH, Moon RE. Decompression sickness and arterial gas embolism. N Engl J Med. 2022;386(13):1254–64.
346. Pollock NW, Buteau D. Updates in decompression illness. Emerg Med Clin. 2017;35(2):301–19.
347. Wilmshurst P, Bryson P. Relationship between the clinical features of neurological decompression illness and its causes. Clin Sci. 2000;99(1):65–75.
348. Russi EW. Diving and the risk of barotrauma. Thorax. 1998;53(suppl 2):S20–4.
349. Alea K. Identifying the subtle presentation of decompression sickness. Aerosp Med Hum Perform. 2015;86(12):1058–62.
350. Weathersby P, et al. Role of oxygen in the production of human decompression sickness. J Appl Physiol. 1987;63(6):2380–7.
351. Moon R. Hyperbaric oxygen treatment for decompression sickness. Undersea Hyperb Med. 2014;41(2):151–7.
352. Blatteau J-E, et al. Prognostic factors of spinal cord decompression sickness in recreational diving: retrospective and multicentric analysis of 279 cases. Neurocrit Care. 2011;15(1):120–7.
353. Moon RE, Mitchell SJ. Hyperbaric oxygen for decompression sickness. Undersea Hyperb Med. 2021;48(2):195–203.
354. Hadanny A, et al. Delayed recompression for decompression sickness: retrospective analysis. PLoS One. 2015;10(4):e0124919.
355. Barnes RC, I.A.-U. ABIM-IM. Intracranial abscess. Undersea Hyperb Med. 2012;39(3):727.

356. Brook I. Microbiology and treatment of brain abscess. J Clin Neurosci. 2017;38:8–12.
357. Xiao F, et al. Brain abscess: clinical experience and analysis of prognostic factors. Surg Neurol. 2005;63(5):442–9.
358. Maniglia AJ, et al. Intracranial abscesses secondary to nasal, sinus, and orbital infections in adults and children. Arch Otolaryngol Head Neck Surg. 1989;115(12):1424–9.
359. Brouwer MC, et al. Brain abscess. N Engl J Med. 2014;371(5):447–56.
360. Lampl L, Frey G. *Intra-cranial abscess*, in *Handbook on hyperbaric medicine*. Springer. 2006 p. 291–304.
361. Scullen T, et al. Surgical approaches in the treatment of intracranial complications of paranasal sinus disease: a review of the literature. World Neurosurg. 2019;130:24–9.
362. Bodilsen J, et al. Long-term mortality and epilepsy in patients after brain abscess: a nationwide population-based matched cohort study. Clin Infect Dis. 2020;71(11):2825–32.
363. Campioli CC, et al. Bacterial brain abscess: an outline for diagnosis and management. Am J Med. 2021;134(10):1210–7. e2
364. Chen M, et al. Management of brain abscesses: where are we now? Childs Nerv Syst. 2018;34(10):1871–80.
365. Bennett M, et al. Hyperbaric oxygen therapy and utilization in infectious disease. Curr Emerg Hospital Med Rep. 2018;6(3):101–9.
366. Tomoye EO, Moon RE. Hyperbaric oxygen for intracranial abscess. Undersea Hyperb Med. 2021;48(1):97–102.
367. M Das J, Tommeraasen MA, Cooper JS. Hyperbaric evaluation and treatment of intracranial abscess. In: StatPearls. StatPearls Publishing, Treasure Island (FL). 2023. PMID: 29630279
368. Kutlay M, et al. Stereotactic aspiration and antibiotic treatment combined with hyperbaric oxygen therapy in the management of bacterial brain abscesses. Neurosurgery. 2005;57(6):1140–6.
369. Tatar, Selcuk, et al. "Hyperbaric oxygen therapy for five days increases blood-brain barrier permeability." Undersea & Hyperbaric Medicine: Journal of the Undersea and Hyperbaric Medical Society, Inc 44.4. 2017;345–55.
370. Bartek J, et al. Hyperbaric oxygen therapy in spontaneous brain abscess patients: a population-based comparative cohort study. Acta Neurochir. 2016;158(7):1259–67.
371. Bachiller FG-C, Caballer AP, Portal LF. Avascular necrosis of the femoral head after femoral neck fracture. Clin Orthop Relat Res. 2002;399:87–109.
372. Hungerford D. Pathogenesis of ischemic necrosis of the femoral head. Instr Course Lect. 1983;32:252–60.
373. Ueo T, et al. Biomechanical aspects of the development of aseptic necrosis of the femoral head. Arch Orthop Trauma Surg. 1985;104(3):145–9.
374. Chandler FA, Peltier LF. Coronary disease of the hip. Clin Orthop Relat Res (1976–2007). 2001;386:7–10.
375. Malizos KN, et al. Osteonecrosis of the femoral head: etiology, imaging and treatment. Eur J Radiol. 2007;63(1):16–28.
376. Delling G. Pathohistology of femoral head necrosis. Der Orthopade. 2007;36(5):404. 406-8, 410
377. Baig SA, Baig M. Osteonecrosis of the femoral head: etiology, investigations, and management. Cureus. 2018;10:8.
378. Björkman A, et al. Factor V Leiden and prothrombin gene mutation: risk factors for osteonecrosis of the femoral head in adults. Clin Orthop Relat Res. 2004;425:168–72.
379. Arbab D, König DP. Atraumatic femoral head necrosis in adults: epidemiology, etiology, diagnosis and treatment. Dtsch Arztebl Int. 2016;113(3):31.
380. Messmer P, Soklić P, Matter P. Femoral neck fracture and femoral head necrosis. Helv Chir Acta. 1993;59(5–6):993–7.
381. Stevens K, et al. Subchondral fractures in osteonecrosis of the femoral head: comparison of radiography, CT, and MR imaging. Am J Roentgenol. 2003;180(2):363–8.
382. Fordyce M, Solomon L. Early detection of avascular necrosis of the femoral head by MRI. J Bone Joint Surg. 1993;75(3):365–7.

383. Manenti G, et al. The role of imaging in diagnosis and management of femoral head avascular necrosis. Clin Cases Miner Bone Metab. 2015;12(Suppl 1):31.
384. Sultan AA, et al. Classification systems of hip osteonecrosis: an updated review. Int Orthop. 2019;43(5):1089–95.
385. Cohen-Rosenblum A, Cui Q. Osteonecrosis of the femoral head. Orthop Clin N Am. 2019;50(2):139–49.
386. Issa K, et al. Osteonecrosis of the femoral head: the total hip replacement solution. Bone Joint J. 2013;95(11_Supple_A):46–50.
387. Algarni AD, Al Moallem HM. Clinical and radiological outcomes of extracorporeal shock wave therapy in early-stage femoral head osteonecrosis. Adv Orthop. 2018;2018:7410246.
388. Massari L, et al. Biophysical stimulation with pulsed electromagnetic fields in osteonecrosis of the femoral head. JBJS. 2006;88(suppl_3):56–60.
389. Beckmann R, et al. Enoxaparin prevents steroid-related avascular necrosis of the femoral head. Sci World J. 2014;2014
390. Luo R-B, et al. Evidence for using alendronate to treat adult avascular necrosis of the femoral head: a systematic review. Med Sci Monit. 2014;20:2439.
391. Hua K-C, et al. The efficacy and safety of core decompression for the treatment of femoral head necrosis: a systematic review and meta-analysis. J Orthop Surg Res. 2019;14(1):1–14.
392. Shannon BD, Trousdale RT. Femoral osteotomies for avascular necrosis of the femoral head. Clin Orthop Relat Res. 2004;418:34–40.
393. Camporesi EM, et al. Hyperbaric oxygen therapy in femoral head necrosis. J Arthroplast. 2010;25(6):118–23.
394. Reis N, et al. Hyperbaric oxygen therapy as a treatment for stage-I avascular necrosis of the femoral head. J Bone Joint Surg. 2003;85(3):371–5.
395. Vezzani G, et al. Hyperbaric oxygen therapy modulates serum OPG/RANKL in femoral head necrosis patients. J Enzyme Inhib Med Chem. 2017;32(1):707–11.
396. Ververidis AN, et al. Bone marrow edema syndrome/transient osteoporosis of the hip joint and management with the utilization of hyperbaric oxygen therapy. J Orthop. 2020;22:29–32.
397. Vezzani G, et al. *Report on HyperBaric Oxygen (HBO2) in femoral head necrosis*. in *Proceedings of the 10th ECHM Consensus Conference of Hyperbaric Medicine, Lille, France*. 2016.
398. Strauss M, et al. *Femoral head necrosis and hyperbaric oxygen therapy.* Hyperbaric medicine practice. Best Publishing Co, 1999. 912.
399. Ditri L, et al. *Femoral head necrosis*, in *Handbook on hyperbaric medicine*. Springer; 2006 p. 547–552.
400. Schaefer TJ, Tannan SC. *Thermal burns*, in *StatPearls [Internet]*. StatPearls Publishing, 2021.
401. Evers LH, Bhavsar D, Mailänder P. The biology of burn injury. Exp Dermatol. 2010;19(9):777–83.
402. Cleland H. Thermal burns: assessment and acute management in the general practice setting. Aust Fam Physician. 2012;41(6):372–5.
403. Schwacha MG, et al. Impact of thermal injury on wound infiltration and the dermal inflammatory response. J Surg Res. 2010;158(1):112–20.
404. Park H-S, et al. Early pathogenic colonisers of acute burn wounds: a retrospective review. Burns. 2017;43(8):1757–65.
405. Robson MC, et al. Prevention of dermal ischemia after thermal injury. Arch Surg. 1978;113(5):621–5.
406. Demling R, Kramer G, Harms B. Role of thermal injury-induced hypoproteinemia on fluid flux and protein permeability in burned and nonburned tissue. Surgery. 1984;95(2):136–44.
407. Lehnhardt M, et al. A qualitative and quantitative analysis of protein loss in human burn wounds. Burns. 2005;31(2):159–67.
408. Sparkes B. Immunological responses to thermal injury. Burns. 1997;23(2):106–13.
409. Gueugniaud P, et al. Current advances in the initial management of major thermal burns. Intensive Care Med. 2000;26(7):848.

410. Schaefer TJ, Szymanski KD. Burn evaluation and management. In: StatPearls. StatPearls Publishing, Treasure Island (FL); 2023. PMID: 28613492.
411. Williams C. Successful assessment and management of burn injuries. Nurs Stand. 2009;23(32):53–9.
412. Wada J, et al. Oxygen hyperbaric treatment for carbon monoxide poisoning and severe burn in coal mine (Hokutanyubari) gas explosion. Igakunoaymi (Japan). 1965;5(1):53–4.
413. Gruber RPC, et al. Hyperbaric oxygen and pedicle flaps, skin grafts, and burns. Plast Reconstr Surg. 1970;45(1):24–30.
414. Chong SJ, et al. Characterization of early thermal burns and the effects of hyperbaric oxygen treatment: a pilot study. Diving Hyperb Med. 2013;43(3):157–61.
415. Culnan DM, et al. Carbon monoxide and cyanide poisoning in the burned pregnant patient: an indication for hyperbaric oxygen therapy. Ann Plast Surg. 2018;80(3 Suppl 2):S106.
416. Smolle C, et al. The history and development of hyperbaric oxygenation (HBO) in thermal burn injury. Medicina. 2021;57(1):49.
417. Gill AÁ, Bell CN. Hyperbaric oxygen: its uses, mechanisms of action and outcomes. QJM. 2004;97(7):385–95.
418. Villanueva E, et al. Hyperbaric oxygen therapy for thermal burns. Cochrane Database Syst Rev. 2004;2
419. Mathieu D. *Handbook on hyperbaric medicine*. Vol. 27. 2006: Springer.
420. Mathes SJ, Alexander J. Radiation injury. Surg Oncol Clin N Am. 1996;5(4):809–24.
421. Hauer-Jensen M, et al. Radiation damage to the gastrointestinal tract: mechanisms, diagnosis, and management. Curr Opin Support Palliat Care. 2007;1(1):23–9.
422. Shadad AK, et al. Gastrointestinal radiation injury: symptoms, risk factors and mechanisms. World J Gastroenterol: WJG. 2013;19(2):185.
423. Feldmeier JJ, Matos LA. *11. DELAYED RADIATION INJURIES (SOFT TISSUE AND BONY NECROSIS)*. 2003.
424. Pollard JM, Gatti RA. Clinical radiation sensitivity with DNA repair disorders: an overview. Int J Radiat Oncol Biol Phys. 2009;74(5):1323–31.
425. Chon BH, Loeffler JS. The effect of nonmalignant systemic disease on tolerance to radiation therapy. Oncologist. 2002;7(2):136–43.
426. Phan C, et al. Matched-control retrospective study of the acute and late complications in patients with collagen vascular diseases treated with radiation therapy. Cancer J. 2003;9(6):461–6.
427. Wasserman T. *Radioprotective effects of amifostine*. in *Seminars in oncology*. 1999.
428. Rogers LF, Goldstein HM. Roentgen manifestations of radiation injury to the gastrointestinal tract. Gastrointest Radiol. 1977;2(1):281–91.
429. Fleckenstein K, et al. *Using biological markers to predict risk of radiation injury*. in *Seminars in radiation oncology*. Elsevier; 2007.
430. Cooper JS, Hanley ME, Hendriksen S, Robins M. Hyperbaric treatment of delayed radiation injury. In: StatPearls. StatPearls Publishing, Treasure Island (FL). 2023. PMID: 29261879.
431. Feldmeier JJ. Hyperbaric oxygen therapy and delayed radiation injuries (soft tissue and bony necrosis): 2012 update. Undersea Hyperb Med. 2012;39(6):1121.
432. Feldmeier JJ, Hampson N. A systematic review of the literature reporting the application of hyperbaric oxygen prevention and treatment of delayed radiation injuries: an evidence based approach. Undersea Hyperb Med. 2002;29(1):4–30.
433. Marx RE, Johnson R. Problem wounds in oral and maxillofacial surgery: the role of hyperbaric oxygen. Problem wounds: the role of oxygen. 1988:65.
434. Cronje F. A review of the Marx protocols: prevention and management of osteoradionecrosis by combining surgery and hyperbaric oxygen therapy. SADJ. 1998;53(10):469–71.

HBOT in Emergency Procedures

8

8.1 Standard and Emergency Indications for HBOT

It is essential for hyperbaric physician and other hyperbaric oxygen therapy personnel to be familiar with all the relevant procedures and protocols to manage the treatment of patients. Based on the risk assessment for a given indication, either standard operating procedure (SOP) or emergency procedure (EP) may be followed which should be properly documented and revised by the hyperbaric facility.

Emergency indications involve serious clinical conditions which require immediate medical attention. Administration of HBOT has the potential to affect the pathophysiological manifestations of these emergent conditions by direct alleviation of clinical symptoms. The treatment plan involves several sessions of HBOT within the first 24 hours in order to stabilize the condition of the patient [1].

The approach of administrating HBOT for emergency conditions may be subjected to technical constraints within hyperbaric facilities as well as condition of patients and availability of medical equipment and trained personnel in the hyperbaric facility. Some of the emergency procedures handled by the administration of HBOT can be listed as mentioned below. However, there may be few inclusions or exclusions based on the risk assessment of patients or capability of hyperbaric facility [2] (Table 8.1).

Training should be provided to all the personnel in a simulated environment in order to develop decision-making capability in a given emergency situation. Safety protocols should be placed in accordance with the operating procedures and utmost care and precautions must be taken to avoid any accident or hazard in the hyperbaric environment.

M. Gupta, I. Somasundaram, *Hyperbaric Oxygen Therapy: Principles and Applications*, https://doi.org/10.1007/978-981-99-7278-4_8

Table 8.1 List of potential emergency procedures handled by HBOT

Emergent condition
• Seizure
• DCS
• Gas embolism
• Pulmonary barotrauma
• Pneumothorax
• Ear or sinus barotrauma
• CRAO
• Cardiac or pulmonary arrest
• Claustrophobia
• Burn injuries
• Accidental extubation
• Carbon monoxide poisoning
• Loss of life support system
• Aggressive patient behavior

8.2 Potential Benefits of HBOT in Emergency Conditions

Subjecting an individual to a hyperbaric environment triggers numerous physiological and biochemical functions beneficial in maintaining homeostasis. Ischemia and tissue necrosis may arise in situations such as burn injuries or crush injuries due to arterial insufficiency and vascular damage. HBOT helps in the maintenance of tissue viability and aids in their survival by ensuring optimum oxygen supply. Additionally, it helps in reducing edema and inflammation at the site of the injury. HBOT is effective to restore oxygen supply to an extensive affected area in case of injury as well as in smaller focal applications such as CRAO. The function of retinal tissues is greatly affected by oxygen availability and requires adequate oxygen supply. HBOT enables replenishment of oxygen supply to ischemic retinal tissues in CRAO and helps in recanalization of central retinal artery [3]. The emergent effects of HBOT are evident in emergency conditions such as carbon monoxide poisoning and severe anemia as overall dissolved oxygen in various body fluids becomes elevated. Further indirect effects of HBOT are manifested by the activation of various protective mechanisms by promoting anti-oxidant function, anti-inflammatory effect, and immune activation. In case of CO poisoning, HBOT aids in rapid oxygenation of Hb by dissociating CO-Hb complex and forming HbO within a short time. High oxygen tension environment is toxic to many pathogenic microorganisms. Emergent conditions associated with a risk of infection derive benefit from HBOT administration as hyperoxic conditions curb down the growth of microorganisms and minimize the levels of toxins. Furthermore, high oxygen tension in blood has been shown to have synergistic role with many commonly used antibiotics.

References

1. Kirby JP. Hyperbaric oxygen therapy emergencies. Mo Med. 2019;116(3):180.
2. Edition T. Hyperbaric oxygen therapy indications.
3. Butler FK, et al. Management of central retinal artery occlusion following successful hyperbaric oxygen therapy: case report. Undersea Hyperb Med. 2018;45(1):101–7.

9 Patient Selection and Patient Care

9.1 Patient Evaluation Before HBOT

According to Undersea and Hyperbaric Medical Society (UHMS), there are 14 indications that are approved for treatment using HBOT. These include both urgent and non-urgent indications. In order to establish the candidature of a patient for HBOT, a thorough examination is required along with the assessment of their medical history. Comprehensive examination prior to starting hyperbaric treatment is essential to review and manage the possibility of any absolute or relative contraindications. Pneumothorax is one of the major absolute contraindications to hyperbaric treatment which need to be managed after proper assessment of risk to benefit ratio. Evaluation of lung imaging and application of chest tube before HBOT is considered as an appropriate measure to manage the condition of pneumothorax. There are several other relative contraindications which need to be addressed before the initiation of HBOT. Following are some of the examples of such relative contraindications:

- Recent thoracic surgery/cataract or retinal surgery
- Obstructive lung disease/asthma
- Chronic sinus condition
- Claustrophobia
- Fever
- History of seizures
- Use of contact lenses
- Untreated cancer
- Diabetes mellitus
- Congestive heart failure.

Additionally, the treatment plan using HBOT must be planned considering the pharmacological interactions with hyperbaric environment. In case of any

M. Gupta, I. Somasundaram, *Hyperbaric Oxygen Therapy: Principles and Applications*, https://doi.org/10.1007/978-981-99-7278-4_9

pharmacological interaction with HBOT, medications must be discontinued before the treatment. HBOT protocols and treatment plans should be designed according to the physical evaluation of the patient to obtain maximum benefit and to ensure minimum risk [1].

9.2 Patient Selection Criteria

Potential candidates to receive HBOT include individuals presented with indications approved for adjunct treatment with prophylactic hyperbaric oxygen or individuals enrolled as control subjects in therapeutic trials for HBOT. Healthcare professionals working in the facility providing HBOT must be aware of the indications appropriate in order to refer patients. Counseling and consultation should be provided to the patients regarding the applications, limitations, and safety regulations with respect to hyperbaric chamber. The effectiveness of HBOT is dependent upon the early prognosis and assessment of any given recommended indication and appropriateness of using HBOT for its management. Thus, the potential risk and benefits of using HBOT for a given patient should be evaluated with the help of appropriate diagnostic parameters as well as patient history [1, 2].

Transcutaneous oxygen pressure ($TcPO_2$) has been considered as an important parameter in the selection of patients for HBOT. $TcPO_2$ measurements at three different sites are taken under normobaric conditions. In selecting the HBOT candidates for treatment of wounds, the $TcPO_2$ at the wound site is recorded. The $TcPO_2$ at the wound site is usually lesser than the reference due to hypoxic conditions. The patient is subjected to hyperbaric condition by providing 100% oxygen at 2.5 ATA. HBOT administration must lead to the normalization of $TcPO_2$ at the wound site indicating its possible therapeutic effect. In case there is no change or insignificant change in the $TcPO_2$ after initial HBOT administration, utilization of HBOT is not justified for the treatment of patient. $TcPO_2$ provides a useful measure in patient selection for indications where localized oxygen pressure is essential for the effectiveness of treatment, e.g. diabetic foot lesions, skin grafts or flaps, problems wounds, skin ulcer, etc. [3].

Assessment of risk factors must be carefully carried out pertaining to the use of HBOT as an adjunct treatment. Monitoring of vital signs of the patient, i.e. blood pressure measurement, pulse rate, respiratory rate, and body temperature helps the physician in formulating a risk vs benefit profile of the patient. Similarly, monitoring blood glucose levels of diabetic patients is essential prior to HBOT as well as during the procedure to avoid any complications. Other factors which need to be assessed while deciding the candidature of patient for HBOT include pulmonary function assessment, assessment of CNS oxygen toxicity, and assessment of cardiac functions. Also, if the patient is febrile before the initiation of HBOT, it may lead to complications in the form of oxygen neurotoxicity and pulmonary toxicity. In such case, medications should be provided to reduce fever. There are various other concerns which may pose risk during the course of hyperbaric treatment. There may be complications related to possible barotrauma such as otitic, sinus, dental, or pulmonary barotrauma. Conversely, administration of high oxygen pressure may lead to oxygen toxicity related complications. Thus, potential HBOT patient must be subjected to mandatory investigations in the form of chest X-ray radiography, spirometry, CT scan (if required), and complete otorhinolaryngological examination in order to rule out any possibility of contraindication [4]. Apart from these, other routine laboratory examinations like lipid profile, ECG, hemoglobin, blood glucose, etc. need to be performed before enrolling the patient for the procedure of HBOT.

References

1. DuBose KJ, Cooper JS. Hyperbaric patient selection. 2018.
2. Yisak Alemayehu R, Kiwanuka F. Hyperbaric oxygen therapy: indications, benefits and nursing management. Int J Caring Sci. 2019;12(1):567–71.
3. Svalestad J, et al. Effect of hyperbaric oxygen treatment on oxygen tension and vascular capacity in irradiated skin and mucosa. Int J Oral Maxillofac Surg. 2014;43(1):107–12.
4. Dey D, et al. The rationale of carrying out specific investigations prior to hyperbaric oxygen therapy. Indian J Aerosp Med. 2013;57(1):56–65.

HBOT in Aging and Regeneration 10

10.1 Aging and Regeneration

The process of aging is complex and multi-factorial which leads to progressive deterioration of physiological processes and reduced efficiency of critical organs. There have been many proposed theories each focused on different factors to explain the advent of aging [1]. However, aging process is more likely a cumulative effect of all such factors. The interplay of these factors manifests as reduced efficiency of signaling pathways, energy transduction mechanisms, weakened musculoskeletal system, reduced skin elasticity, and several other signs of aging (Fig. 10.1) [2]. Programmed theory and damage and error theory are the two broad categories of theories of aging. Programmed theory is based on the assumption that aging is a process governed by the inherent biological clock of an organism similar to all other biological events in their life span. It is further categorized into theory of programmed longevity, endocrine theory, and immunological theory of aging [3]. Conversely, damage and error theory takes into account the role of external factors in culminating the effects of aging. It is also divided into sub-categories as wear and tear theory, rate of living theory, cross-linking theory, free-radical theory, and somatic DNA damage theory [4].

The common denominator of all the theories of aging essentially involves the progressive loss of ability of an individual to maintain homeostasis. There is an increased rate of mutations, excessive oxidative damage due to mitochondrial dysfunction, loss of connective tissues, loss of elasticity, and muscular atrophy. All these detrimental effects arise due to an imbalance between the anabolic and catabolic functions of body [5–7]. The repair and regenerative capacity of cells get reduced with age throughout the organs which affects the function of vital organs such as heart, brain, liver, muscles, lungs, and kidneys. During the lifetime of an individual, all the cells except neurons are constantly regenerated to replace the senescent cells. The restriction in this regenerative ability with age occurs as the cells have limitations to re-entering cell cycle [8]. The regenerative process is

M. Gupta, I. Somasundaram, *Hyperbaric Oxygen Therapy: Principles and Applications*, https://doi.org/10.1007/978-981-99-7278-4_10

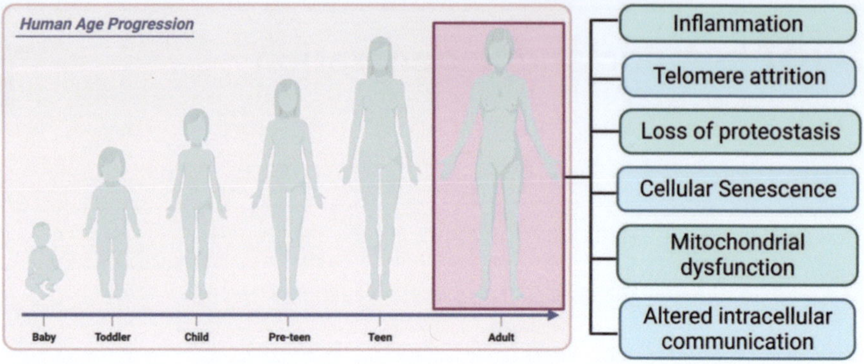

Fig. 10.1 Progression of age and age-associated changes. Human age progression involves multiple physiological changes throughout the life cycle of an individual in each stage of life. As the age progresses, multiple factors contribute to the decline in overall health due to a net change toward degenerative bodily processes

activated upon cellular injury and orchestrated by the release of various systemic factors. Regenerative therapies are designed to enhance the organ and tissue specific components which regulate the repair and regeneration process. Current scenario of research in aging has been focused on understanding the molecular mechanisms of aging and the factors which govern the process as well as to enhance the regenerative capability of cells and tissues.

10.2 HBOT as a Regenerative and Anti-Aging Therapy

HBOT has been implemented in a large number of clinical conditions where it is found to be highly beneficial. HBOT enables increased bioavailability of oxygen throughout the tissues by elevating the partial pressure of oxygen in the blood. The therapeutic effects of HBOT arise from the direct effect of oxygen in preservation of oxidative environment and indirectly by modulating a wide array of physiological functions (Fig. 10.2) [9]. The higher partial pressure of oxygen triggers the release of various regulatory factors involved in physiological functions such as immunological response, anti-oxidant functions, regeneration, and repair process [10–12]. The effects of HBOT in wound healing and tissue regeneration have been well studied [13, 14]. Its role in promoting cell viability and regeneration capacity has drawn attention toward its potential use as a regenerative therapy and anti-aging therapy.

In the cellular level, cell senescence and telomere shortening are the two major factors responsible for limitations of cellular regeneration [15]. As the cell complete its life span, it becomes senescent, i.e. destined to go through programmed cell death via apoptosis. The rate of cellular senescence becomes higher with age due to accumulation of genetic mutations and shortening of telomeres which arrests its further propagation into nascent cells [16]. Telomeres are non-coding sequences of

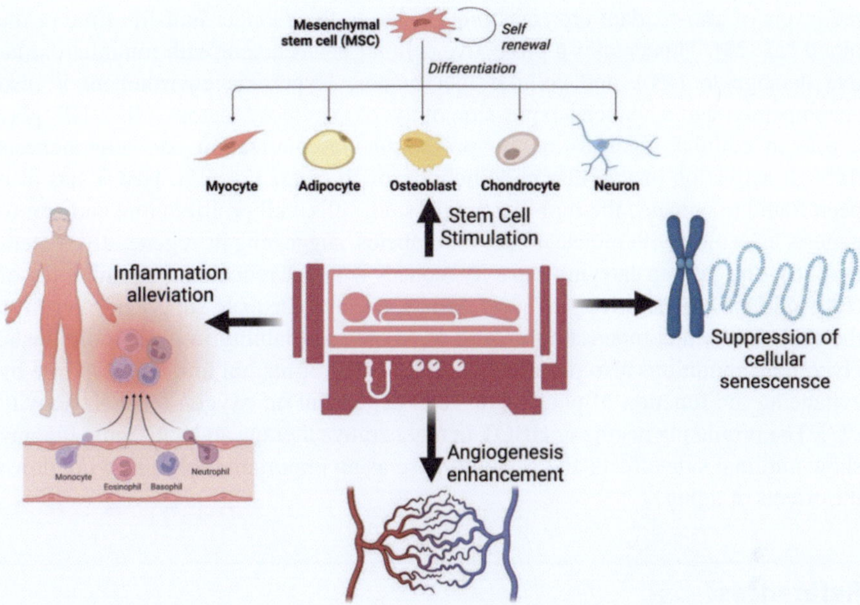

Fig. 10.2 Effect of HBOT in aging and regeneration. HBOT can be beneficial as an anti-aging therapy as it affects the functions directly associated with aging like suppressing inflammatory pathways, activating stem cells renewal, enhancing telomerase activity, and promoting angiogenesis

DNA which are important in safeguarding genetic integrity by protecting the chromosomal DNA from exonucleases. Telomerase is a ribozyme which functions in replenishing telomeres after each successive cell division. Since most somatic cells lack telomerase, the length of telomeres gets shortened over time leading ultimately to their programmed cell death. Telomeres shorten at a rate of 20–40 bases per year during the lifetime and as the age progresses, a higher number of senescent cells get accumulated [15, 17, 18]. Extrinsic factors like lifestyle, food habits, pollution, etc. affect the rate of telomere shortening [19, 20]. Conversely, rate of telomeric shortening can be reduced by using therapeutic measures. HBOT has been studied for its effect on rate of telomere shortening in peripheral blood mononuclear cells (PBMCs). There was a decline in the number of senescent cells by 10–37% upon exposure to HBOT. In an aging population of blood cells, there was 20% increase in the telomere length [21]. Intermittent HBOT exposure resulted in increase in telomere length, decreased rate of telomere shortening, and enhanced cellular lifetime. Similar effects have been observed in case of aerobic exercise where telomere length was found to increase by 5% in a duration of 6 months of aerobic endurance training [22]. The effects may be attributed to the hyperoxic–hypoxic paradox manifesting due to intermittent hyperoxic conditions [22]. Initial hyperoxic conditions result in an increased ROS production which in turn leads to overexpression of antioxidant genes such as superoxide dismutase (SOD), glutathione peroxidase (GPx), and glutathione reductase (GRx). Eventually, the levels of ROS drop rapidly, while the levels of anti-oxidant enzymes prevail due to their longer half-life time in the blood [22, 23]. This creates a protective cellular environment with minimal oxidative damage to DNA and cellular membranes. Hyperoxic environment is also accompanied with enhanced expression of hypoxia induced factors (HIF). HIF play a role in cellular regeneration by promoting vascularization and angiogenesis through activation of vascular endothelial growth factor [24, 25]. HBOT has also been found to enhance the mobilization of stem cells, cell proliferation, and angiogenesis in a study with mouse model for diabetes, suggesting its regenerative potential [26]. The oxygen carrying capacity reduces with age reducing the availability of oxygen in the regenerative processes. HBOT ensures adequate supply of oxygen to the tissues which compensate the lack of oxygen availability in older individuals. Hyperoxic conditions also promote the functioning of natural immune response by enhancing the function of phagocytic cells dependent on oxygen for their activity [27]. The promising results of HBOT in regenerative therapy and anti-aging therapy show minimal side effects and it would pose as an important tool in slowing down the effects of aging.

References

1. Bengtson VL, Settersten R Jr. Handbook of theories of aging. Cham: Springer; 2016.
2. Franceschi C, et al. The network and the remodeling theories of aging: historical background and new perspectives. Exp Gerontol. 2000;35(6–7):879–96.
3. Mitteldorf J. Programmed and non-programmed theories of aging. Russ J Gen Chem. 2010;80(7):1465–75.
4. Jin K. Modern biological theories of aging. Aging Dis. 2010;1(2):72.
5. Kong Y, Trabucco SE, Zhang H. Oxidative stress, mitochondrial dysfunction and the mitochondria theory of aging. Aging. 2014;39:86–107.
6. Tu Y, Quan T. Oxidative stress and human skin connective tissue aging. Cosmetics. 2016;3(3):28.
7. Muscaritoli M, et al. Muscle atrophy in aging and chronic diseases: is it sarcopenia or cachexia? Intern Emerg Med. 2013;8(7):553–60.
8. Khokhlov A. Impairment of regeneration in aging: appropriateness or stochastics? Biogerontology. 2013;14(6):703–8.
9. Sen S, Sen S. Therapeutic effects of hyperbaric oxygen: integrated review. Med Gas Res. 2021;11(1):30.
10. De Wolde SD, et al. The effect of hyperbaric oxygen therapy on markers of oxidative stress and the immune response in healthy volunteers. Front Physiol. 2022;13:72.
11. De Wolde SD, et al. The effects of hyperbaric oxygenation on oxidative stress, inflammation and angiogenesis. Biomol Ther. 2021;11(8):1210.
12. Edwards ML. Hyperbaric oxygen therapy. Part 2: application in disease. J Vet Emerg Crit Care. 2010;20(3):289–97.
13. Lindenmann J, et al. Hyperbaric oxygen therapy and tissue regeneration: a literature survey. Biomedicine. 2022;10(12):3145.
14. Thackham JA, McElwain DS, Long RJ. The use of hyperbaric oxygen therapy to treat chronic wounds: a review. Wound Repair Regen. 2008;16(3):321–30.
15. Zhu Y, et al. Telomere and its role in the aging pathways: telomere shortening, cell senescence and mitochondria dysfunction. Biogerontology. 2019;20(1):1–16.
16. Muñoz-Espín D, Serrano M. Cellular senescence: from physiology to pathology. Nat Rev Mol Cell Biol. 2014;15(7):482–96.

17. Rubelj I, Vondracek Z. Stochastic mechanism of cellular aging—abrupt telomere shortening as a model for stochastic nature of cellular aging. J Theor Biol. 1999;197(4):425–38.
18. Fathi E, et al. Telomere shortening as a hallmark of stem cell senescence. Stem Cell Invest. 2019;6:6.
19. Hou L, et al. Air pollution exposure and telomere length in highly exposed subjects in Beijing, China: a repeated-measure study. Environ Int. 2012;48:71–7.
20. Todendi PF, et al. Biochemical profile, eating habits, and telomere length among Brazilian children and adolescents. Nutrition. 2020;71:110645.
21. Hachmo Y, et al. Hyperbaric oxygen therapy increases telomere length and decreases immunosenescence in isolated blood cells: a prospective trial. Aging (Albany NY). 2020;12(22):22445.
22. Werner CM, et al. Differential effects of endurance, interval, and resistance training on telomerase activity and telomere length in a randomized, controlled study. Eur Heart J. 2019;40(1):34–46.
23. Burtscher J, et al. Adaptive responses to hypoxia and/or hyperoxia in humans. Antioxid Redox Signal. 2022;37(13):887–912.
24. Schoppmann SF, et al. Hypoxia inducible factor-1α correlates with VEGF-C expression and lymphangiogenesis in breast cancer. Breast Cancer Res Treat. 2006;99(2):135–41.
25. Yamamoto N, et al. VEGF and bFGF induction by nitric oxide is associated with hyperbaric oxygen-induced angiogenesis and muscle regeneration. Sci Rep. 2020;10(1):1–13.
26. Peña-Villalobos I, et al. Hyperbaric oxygen increases stem cell proliferation, angiogenesis and wound-healing ability of WJ-MSCs in diabetic mice. Front Physiol. 2018;9:995.
27. Grim PS, et al. Hyperbaric oxygen therapy. JAMA. 1990;263(16):2216–20.

Oxygen Toxicity and Side Effects of HBOT

11

11.1 Oxygen Toxicity

The toxicity of oxygen was first documented by Paul Bert in 1878 through his experiments on the effects of hyperbaric oxygen on central nervous system. Oxygen toxicity affects the CNS resulting in manifestation as seizures which is referred as "Paul Bert effect" [1]. Apart from the detrimental effects on CNS, the toxic effects of oxygen were also evident on pulmonary system. The pulmonary toxicity of oxygen was first reported by Lorraine Smith and is known as "Lorraine Smith effect" [2]. Most of the effects of oxygen toxicity are reversible but prolonged exposure to pure oxygen or continuous administration of oxygen under hyperbaric conditions may lead to irreversible neurological damage or even death. Extreme detrimental effects of oxygen toxicity were first discovered by John Bean in 1945 and named as "John Bean effect" [3].

11.2 Mechanism of Oxygen Toxicity

The primary basis of oxygen toxicity involves the damage caused by the excessive generation of free radical species. The mitochondrial function in the generation of cellular energy depends on oxygen as the final electron acceptor in the electron transport chain [4]. Free radicals in the form of superoxide anion are formed in the cells under normal conditions which are scavenged by anti-oxidant enzymes. However, under hyperbaric conditions, the rate of superoxide free radical generation is markedly increased leading to formation of other reaction products of oxygen metabolism such as hydrogen peroxide, singlet oxygen species, and hydroxyl radicals [5]. The reactive oxygen species can act on membrane lipids, DNA, and enzymes leading to detrimental effects. There is an overall disturbance in the cellular integrity due to shift in cellular redox potential of glutathione in oxidative

M. Gupta, I. Somasundaram, *Hyperbaric Oxygen Therapy: Principles and Applications*, https://doi.org/10.1007/978-981-99-7278-4_11

state, DNA damage, enzyme inactivation of enzymes with sulfhydryl group (-SH) active site, and damage to biomembrane [6–8].

11.3 Physiological Manifestations of Oxygen Toxicity

The ill-effects of oxygen toxicity are manifested mainly in the form of pulmonary oxygen toxicity, CNS toxicity, and retinopathy [9]. The limits of safe administration of HBOT in humans have been determined to be 3 h at 3 ATA and 30–40 min at 4 ATA [10]. Prolonged exposure to hyperoxic conditions leads to the manifestation of clinical symptoms of oxygen toxicity. The nervous system is affected upon exposure to 3 ATA pressure, while the disruptions in pulmonary functions begin under 2 ATA pressure for long time period [11]. Lorraine Smith documented the toxic effect of oxygen on pulmonary system in 1899. He described that the toxic effect of oxygen is dependent on the partial pressure (pO_2) and the effects become more prominent as pO_2 increases. Pulmonary oxygen toxicity is observed upon oxygen inspired at 0.45 ATA to 1.6 ATA, while CNS oxygen toxicity begins manifesting at pressures above 1.6 ATA [12].

Breathing oxygen at hyperbaric pressure can cause acute lung injury due to the action of ROS and increase in pulmonary vascular pressure leading to barotrauma [13]. Hyperoxic conditions stimulate the sympathetic nervous system which in turn causes depression in the left ventricular function. This results in acute left atrial and pulmonary hypertension. One of the major factors that deteriorate the pulmonary function under hyperbaric conditions is the occurrence of alveolar and interstitial edema [14]. Hyperoxia also activates type II epithelial cells and fibroblasts in lungs leading to collagen deposits and fibrosis [15]. Pulmonary oxygen toxicity depends on the duration of exposure as the fibrosis and emphysema in lungs are not reverted to normal condition even after the discontinuation of hyperbaric conditions. Serial pulmonary function studies are employed to monitor the progression of pulmonary toxicity. Unit pulmonary toxic dose (UPTD) is used to determine the toxic dose of oxygen in the treatment protocols. UPTD is an arbitrary measure to express the toxic effect of oxygen in terms of equivalent exposure of oxygen at 1 ATA pressure [16].

The manifestation of CNS oxygen toxicity in humans occurs at a higher oxygen pressure as compared to the pressure required to cause pulmonary oxygen toxicity. The most common clinical feature of CNS oxygen toxicity is the occurrence of seizures. The hyperoxia related seizures are reversible in nature and cause no permanent neurological damage. The clinical symptoms of hyperoxia induced CNS toxicity are alleviated upon subjecting to normobaric pressure. The symptoms appear in the form of minor neurological disturbances such as nausea, twitching of eyes and lips, headache, dizziness, blurred vision, or tunnel vision [17]. Neurological manifestations of CNS oxygen toxicity arise due to decrease in excitability of neurons and hampering in nerve impulse conduction. There is an increase in the spontaneous release of neurotransmitters and reduction in inhibitory nerve transmission at the neuromuscular junction leading to seizures and convulsions [9, 18]. Hyperbaric conditions affect the release of neurotransmitters such as GABA, acetylcholine, glutamate, dopamine, and NO. Prolonged exposure to hyperbaric conditions results in NO overproduction and increased oxygen delivery to brain due to cessation of vasoconstriction under the effect of NO [19]. Administration of NMDA antagonists MK-801 or NOS inhibitor L-N (G)-nitro-arginine methyl ester has been found to be protective against hyperoxia induced neurological problems [20, 21].

There are several factors which are found to be responsible for enhancing the toxic effects of oxygen. Administration of intravenous perfluorocarbon along with supplemental oxygen enhances oxygen delivery to brain [22, 23]. Higher CO_2 concentration is a risk factor for oxygen toxicity as it can enhance the production of ROS and promote cellular stress [24]. Mild hyperthermic conditions can also enhance oxygen toxicity by promoting increased uptake of oxygen by tissues [25]. Similarly higher humidity has been considered as risk factor for oxygen toxicity [26]. In order to safely administer HBOT for therapeutical purpose, assessment of risks and benefits needs to be carefully performed by the practitioner. Use of free radical scavengers and anti-oxidants is proved to be beneficial in minimizing the free radical induced cellular damage. Intermittent HBOT exposure within the safe limits also helps in reducing the risk of toxicity by allowing time for recovery from harmful oxygen induced physiological changes.

11.4 Side Effects of HBOT

HBOT finds its use as an adjunctive therapy in multiple pathological conditions. The administration of HBOT is carried out by following strict treatment protocols in order to gain maximum benefits. However, as with any medical procedure, HBOT administration can have potential risks and side effects which need to be carefully considered while planning treatment strategy. Most of the side effects of HBOT are mild and reversible in nature but there is a potential risk to the patients with pre-existing medical conditions. The mechanism of HBOT depends on the direct effect of pressure and hyperoxic environment as well as indirect effects with respect to modulation of various physiological functions. Although these direct or indirect effects are of significance from therapeutic point of view, adverse effects may be experienced by the patients undergoing hyperbaric treatment.

The increased partial pressure of oxygen affects the air cavities in the body which may lead to potential side effects. Sudden changes in pressure during the phase of compression or decompression give rise to a condition called barotrauma. Different types of barotraumas are distinguished based on the affected site as follows.

11.4.1 Pulmonary Barotrauma

Change in the volume of air in the closed body cavity occurs under the influence of pressure. Lungs are the cavities with open system and do not show any adverse effect of increased barometric pressure under normal conditions. However, in case of pre-existing lung pathology such as asthma, bullous lung disease, or chronic obstructive pulmonary disease, there is a risk of pulmonary barotrauma [27, 28]. The expansion of gas during decompression leads to disruption of lung parenchyma and manifestations of intrapulmonary hemorrhage, simple pneumothorax, and tension pneumothorax. Pulmonary pneumothorax can be potentially dangerous especially if it leads to occurrence of tension pneumothorax from simple pneumothorax. Cardiovascular collapse can be caused under the condition of tension pneumothorax leading to death. In such cases, thoracostomy tube insertion needs to be performed in order to relieve the tension pneumothorax [29, 30]. The patients scheduled to undergo hyperbaric treatment are thoroughly screened for the presence of any pulmonary disease which could lead to the development of tension pneumothorax. The HBOT protocols need to be designed involving slower decompression rate for such patients to minimize the risk of pulmonary barotrauma.

11.4.2 Middle Ear Barotrauma

Middle ear barotrauma (MEB) is the most common side effect experienced by patients during the administration of HBOT [31]. MEB occurs during the compression phase of HBOT resulting in ear pain, problems related to ear equalization, general discomfort in ear, and feeling of pressure. Although common, MEB can lead to middle ear edema or rupture of tympanic membrane causing hearing impairment. In some cases, MEB can progress toward inner ear and may result in oval window membrane rupture and sensorineural hearing loss [32]. The risk of occurrence of MEB depends on the rate of compression during the procedure of HBOT.

11.4.3 Sinus/Paranasal Barotrauma

Sinus and paranasal barotrauma may occur during HBOT procedure and is characterized by sensation of pressure in the frontal sinus during the compression phase leading to a condition called barosinusitis. Occlusion of sinus takes place during compression due to formation of a negative pressure gradient causing inflammation of sinus mucosal surface [33, 34]. Inflammatory conditions lead to pain and edema which further obstructs and traps the air in the sinus cavity during decompression phase. Sinus/paranasal barotrauma is experienced by patients with incidence of upper respiratory tract infection, mucociliary dysfunction, or allergic rhinitis. In order to control the severity of sinus/paranasal barotrauma, nasal decongestant and antihistamines are usually administered before HBOT.

11.4.4 Dental Barotrauma

The pressure changes during both compression and decompression phase of HBOT affect the dental cavity causing a sensation of pain in tooth. There is an expansion of air within the dental cavity during decompression causing stimulation of nociceptors in the maxillary sinuses. The dental stress caused by expanding air bubbles may cause dental fracture which is also termed as odontecrexis [35, 36]. Presence of dental infections or different expansion manner of dental enamel and dental pulp makes the patient more susceptible to dental barotrauma. Precautionary measures to avoid dental barotrauma should be taken in the form of thorough dental examination and treatment of any dental infection prior to the administration of HBOT.

11.4.5 Other Side Effects of HBOT

HBOT administration results in an increase in the production of ROS which may affect various tissues in the body. Especially tissues sensitive to cellular stress such as ocular tissues are more sensitive to ROS and lead to manifestations of ocular side effects such as hyperoxic myopia, cataracts, or retrolental fibroplasia [37]. Increased partial pressure of oxygen affects vascular system resulting in irregularity in blood pressure. Patients may also undergo mild hypoglycemia due to increased metabolic rate and utilization of blood glucose under hyperoxic conditions [33]. Claustrophobia may be experienced by some patients due to enclosed space of hyperbaric chamber which needs to be addressed before the hyperbaric treatment [38].

11.5 Specific Drug Interaction with HBOT

Oxygen itself functions as a drug and can have potential interaction with many commonly used drugs. The effect of drugs may either be reduced or enhanced based on the behavior of drug under hyperbaric and hyperoxic condition. Antibiotics shows synergistic effect with oxygen as the production of ROS during HBOT enhances the microbial killing. Similarly, carbapenem drugs such as imipenem, which is used against *P. aeruginosa,* show enhanced effectiveness under hyperoxic conditions. The production of superoxide ions by macrophages is elevated during HBOT promoting the eradication of *P. aeruginosa* infection in burn wounds [39]. Hypoxia is a hallmark of cancerous tissues and the drug response is affected by its local environment. The effect of anti-cancer drugs is modulated during HBOT due to elevated oxygen supply to cancerous tissues. In general, the effect of anti-cancer drugs gets enhanced under hyperoxic conditions. Cytotoxicity in cancer cells has been found to be reduced with simultaneous treatment with adriamycin and HBOT, while treatment with adriamycin after or before HBOT enhances cytotoxicity. The chemotherapeutic effects of doxorubicin were also found to be increased when conjugated with HBOT [40, 41]. However, there may be a risk of cardiotoxicity with its simultaneous use with HBOT. Similarly, potentiation of sorafenib has been shown in inhibiting the growth and promotion of apoptosis in case of hepatocellular carcinoma [42]. Cardiovascular drugs such as β-blockers, adrenomimetics, and ganglion blockers show hypotensive effect when administered along with HBOT, while cardiac glycoside drugs like digoxin exhibit reduced effect with HBOT [43, 44]. Drugs acting on CNS mostly show unfavorable effects in conjunction with HBOT. CNS stimulants like amphetamines and narcotic drugs result in depressed respiration and increase in incidence of oxygen toxicity. There are other drugs which are known to increase oxygen toxicity such as acetazolamide which prevents vasoconstriction and promotes oxidative damage to neuronal tissues. Disulfiram promotes superoxide ions mediated tissue damage by inhibiting superoxide dismutase [11]. Conversely, therapeutic interventions that enhance the action of anti-oxidants are helpful in reducing the oxygen toxicity. Supplementation of vitamin E, magnesium, ergot derivatives during the procedure of HBOT is beneficial in protection against oxidative damage [45].

11.6 Effect of HBOT on Drug Mechanism

Pharmacological active compounds interact with the biological system to impart their effect. There are various factors which can affect the functions of these drugs and can alter their therapeutic action. HBOT can potentially interact with certain drugs resulting in the modulation of their efficiency. Oxygen can itself be considered as a pharmacological agent as it is known to show therapeutic effects in the form of reducing edema, promoting vasoconstriction, modulating immunological functions and other effects. Most patients undergoing hyperbaric treatment are also prescribed a drug regimen for their existing clinical condition. HBOT can either reduce or potentiate the effects of these drugs depending on their interaction. Also, some drugs can give rise to side effects under the influence of hyperbaric and hyperoxic conditions. HBOT may affect the absorption rate of drug by the tissues as well as their pharmacodynamics and pharmacokinetics [46]. The rate of blood flow and cardiac output is reduced under hyperbaric and hyperoxic conditions which also reduces the rate of drug absorption. Drug clearance from the body is dependent upon the systemic blood flow in liver and activity of hepatic enzymes. Reduced rate of hepatic blood perfusion decreases the rate of drug clearance from liver. Similarly, decreased renal perfusion rate and glomerular filtration rate also reduce the drug elimination and excretion through kidneys [47]. Additionally, the increase in ROS occurs in cells and tissues during hyperbaric treatment. ROS induced lipid peroxidation and alteration of membrane proteins can potentially modulate the mechanism of drug action as protein-drug binding and membrane permeability of drugs can be reduced [33]. Hyperbaric practitioners need to consider the potential side effects and interaction of drugs with HBOT in order to avoid any harmful effect during the procedure.

References

1. Wolfe WG, DeVries WC. Oxygen toxicity. Annu Rev Med. 1975;26(1):203–17.
2. Smith JL. The pathological effects due to increase of oxygen tension in the air breathed. J Physiol. 1899;24(1):19.

3. Bean JW. Effects of oxygen at increased pressure. Physiol Rev. 1945;25(1):1–147.
4. Atamna H, Mackey J, Dhahbi JM. Mitochondrial pharmacology: electron transport chain bypass as strategies to treat mitochondrial dysfunction. Biofactors. 2012;38(2):158–66.
5. Ross ME. Myopia and free radical generation associated with hyperbaric oxygen therapy. Udaipur: Pacific University; 1995.
6. Rice-Evans CA, Burdon RH. Free radical damage and its control. Amsterdam: Elsevier; 1994.
7. Recknagel RO, Glende EA, Britton RS. Free radical damage and lipid peroxidation. In: Hepatotoxicology. Boca Raton, FL: CRC; 2020. p. 401–36.
8. Ketterer B. Glutathione S-transferases and prevention of cellular free radical damage. Free Radic Res. 1998;28(6):647–58.
9. Jain KK. Oxygen toxicity. In: Textbook of hyperbaric medicine. Cham: Springer; 2017. p. 49–60.
10. Lam G, et al. Hyperbaric oxygen therapy: exploring the clinical evidence. Adv Skin Wound Care. 2017;30(4):181–90.
11. Plafki C, et al. Complications and side effects of hyperbaric oxygen therapy. Aviat Space Environ Med. 2000;71(2):119–24.
12. Arieli R. Calculated risk of pulmonary and central nervous system oxygen toxicity: a toxicity index derived from the power equation. Diving Hyperb Med. 2019;49(3):154.
13. Rivalland G, Mitchell SJ, Van Schalkwyk JM. Pulmonary barotrauma and cerebral arterial gas embolism during hyperbaric oxygen therapy. Aviat Space Environ Med. 2010;81(9):888–90.
14. Weaver LK, Churchill S. Pulmonary edema associated with hyperbaric oxygen therapy. Chest. 2001;120(4):1407–9.
15. Lang Y-D, et al. The renin-angiotensin system mediates hyperoxia-induced collagen production in human lung fibroblasts. Free Radic Biol Med. 2010;49(1):88–95.
16. Harabin AL, et al. Predicting pulmonary O2 toxicity: a new look at the unit pulmonary toxicity dose. Bethesda, MD: Naval Medical Research Inst; 1985.
17. Arieli R, et al. CNS oxygen toxicity in closed-circuit diving: signs and symptoms before loss of consciousness. Aviat Space Environ Med. 2006;77(11):1153–7.
18. Zhang J, et al. Cerebral amino acid, norepinephrine and nitric oxide metabolism in CNS oxygen toxicity. Brain Res. 1993;606(1):56–62.
19. Oury TD, et al. Extracellular superoxide dismutase, nitric oxide, and central nervous system O2 toxicity. Proc Natl Acad Sci. 1992;89(20):9715–9.
20. Allen BW, Demchenko IT, Piantadosi CA. Two faces of nitric oxide: implications for cellular mechanisms of oxygen toxicity. J Appl Physiol. 2009;106(2):662–7.
21. Chavko M, Braisted J, Harabin A. Effect of MK-801 on seizures induced by exposure to hyperbaric oxygen: comparison with AP-7. Toxicol Appl Pharmacol. 1998;151(2):222–8.
22. Liu S, et al. Perfluorocarbon-facilitated CNS oxygen toxicity in rats: reversal by edaravone. Brain Res. 2012;1471:56–65.
23. Demchenko IT, et al. Brain oxygenation and CNS oxygen toxicity after infusion of perfluorocarbon emulsion. J Appl Physiol. 2012;113(2):224–31.
24. Ezraty B, et al. CO2 exacerbates oxygen toxicity. EMBO Rep. 2011;12(4):321–6.
25. Issels RD, Fink RM, Lengfelder E. Effects of hyperthermic conditions on the reactivity of oxygen radicals. Free Radic Res Commun. 1986;2(1–2):7–18.
26. Lin Y, Jamieson D. Effect of humidity on hyperoxic toxicity. J Appl Physiol. 1993;75(5):1980–3.
27. Hillman K. Pulmonary barotrauma. Clin Anaesthesiol. 1985;3(4):877–98.
28. Kennedy-Little, D. And T. Sharman, Pulmonary barotrauma. StatPearls Publishing, Treasure Island, FL 2020.
29. Barton ED. Tension pneumothorax. Curr Opin Pulm Med. 1999;5(4):269.
30. Chen K-Y, et al. Pneumothorax in the ICU: patient outcomes and prognostic factors. Chest. 2002;122(2):678–83.
31. ONeill OJ, Brett K, Frank AJ. Middle ear barotrauma. Treasure Island, FL: StatPearls; 2018.
32. Pullen FW. Perilymphatic fistula induced by barotrauma. Otol Neurotol. 1992;13(3):270–2.
33. Heyboer M III, et al. Hyperbaric oxygen therapy: side effects defined and quantified. Adv Wound Care. 2017;6(6):210–24.

34. Fitzpatrick D, et al. Risk factors for symptomatic otic and sinus barotrauma in a multiplace hyperbaric chamber. Undersea Hyperb Med. 1999;26(4):243.
35. Zadik Y. Dental barotrauma. Int J Prosthodont. 2009;22(4):354.
36. Rajpal PS, et al. Dental Barotrauma and Barodontalgia. J Med Sci Clin Res. 2014;2:3477–83.
37. McMonnies CW. Hyperbaric oxygen therapy and the possibility of ocular complications or contraindications. Clin Exp Optom. 2015;98(2):122–5.
38. Camporesi EM. Side effects of hyperbaric oxygen therapy. Undersea Hyperb Med. 2014;41(3):253–7.
39. Lima FL, et al. Effects of hyperbaric oxygen on Pseudomonas aeruginosa susceptibility to imipenem and macrophages. Future Microbiol. 2015;10(2):179–89.
40. Monstrey SJ, et al. Hyperbaric oxygen therapy and free radical production: an experimental study in doxorubicin (Adriamycin) extravasation injuries. Ann Plast Surg. 1997;38(2):163–8.
41. Karagoz B, et al. Hyperbaric oxygen therapy does not potentiate doxorubicin-induced cardiotoxicity in rats. Basic Clin Pharmacol Toxicol. 2008;102(3):287–92.
42. Peng H-S, et al. Synergistic inhibitory effect of hyperbaric oxygen combined with sorafenib on hepatoma cells. PLoS One. 2014;9(6):e100814.
43. Filatov A, Reznikov K. Effect of adrenomimetic, adreno-and ganglionic-blocking agents during hyperbaric oxygenation. Farmakol Toksikol. 1982;45(2):35–9.
44. Jain KK. Drug interactions with hyperbaric oxygenation. In: Textbook of hyperbaric medicine. Cham: Springer; 2017. p. 85–9.
45. Dennog C, et al. Antioxidant status in humans after exposure to hyperbaric oxygen. Mutat Res. 1999;428(1–2):83–9.
46. Smith RG. an appraisal of potential drug interactions regarding hyperbaric oxygen therapy and frequently prescribed medications. Wounds. 2011;23(6):147–59.
47. Gandhi J, et al. Clinical utility of hyperbaric oxygen therapy in genitourinary medicine. Med Gas Res. 2018;8(1):29.

Contraindications of HBOT 12

In 1994, European Committee for Hyperbaric Medicine issued recommendations on the functioning of hyperbaric facilities [1, 2]. The hyperbaric facility should be well equipped with respect to technical as well as staff competence to carry out and continue HBOT for the required period of time. Also, proper safety measures should be in place to handle any potential accident or derangement in the course of hyperbaric treatment. The utilization of HBOT for therapeutic purpose depends on careful risk assessment and to minimize complications during the procedure.

12.1 Risk Assessment of Barometric Changes

Sudden changes in the barometric pressure can cause injuries affecting the physiological functions of an individual. Such injuries are termed as barotrauma. HBOT subjects an individual to a large variation in barometric pressure making them susceptible to barotrauma. Barometric pressure changes affect the gas filled cavities of body such as ears, sinuses, gut, and lungs. HBOT involves changes in barometric pressure during both the compression phase and decompression phase. In the compression phase, mainly ear and sinuses are the affected regions if the air entry to these cavities is obstructed. Middle ear barotrauma is the most common complication during the compression phase of HBOT [3]. The incidence of ear barotrauma varies according to the clinical condition of the patient as well as clinical settings, supervision, and variations in compression protocols [4]. In the decompression phase, lung barotrauma is most frequent which may cause pneumomediastinum, hemoptysis, pneumothorax, subcutaneous emphysema, or arterial gas embolism. Decompression can cause

M. Gupta, I. Somasundaram, *Hyperbaric Oxygen Therapy: Principles and Applications*, https://doi.org/10.1007/978-981-99-7278-4_12

development of tension pneumothorax from simple pneumothorax which may lead to cardiovascular dysfunction, respiratory distress, or cardiac arrest. Patients with previous history of lung disorders or cardiac disorders are at high risk of contracting pulmonary barotrauma during the procedure of HBOT. Thus, the clinical history of patient needs to be assessed before administration of HBOT [5–7].

12.2 Risk Assessment of Oxygen Breathing Under Hyperbaric Conditions

Consideration of the adverse effects of oxygen toxicity is important in order to ensure safe administration of HBOT. Oxygen toxicity may lead to complications affecting different organ systems. Patients administered with HBOT intake oxygen at elevated pressures result in hyperoxic conditions. Development of oxygen toxicity may occur as frequent as 1 in 2000 treatment sessions. The duration of HBOT and operating pressure is critical in this case and the risk factor varies accordingly. The risk factor could be as high as 1 in 200 treatment sessions at 2.8–3.0 ATA pressure to as low as 1 in 10,000 treatment sessions at 2.0 ATA pressure [8].

The manifestation of oxygen toxicity has been observed primarily in ocular, pulmonary, or neurological functions [9]. HBOT may cause visual disturbance in some patients in the form of temporary myopia, problems in night vision, or progressive cataract formation. The patients may also experience visual hallucinations and transient unilateral loss of vision. Most of these effects are reversible in nature and return to normalcy after the discontinuation of HBOT. The occurrence of pulmonary toxicity is rare during HBOT as the protocols for treatment are designed to avoid such incidences [10]. Symptoms of pulmonary toxicity may appear after prolonged exposure to hyperoxic conditions which manifests as histological changes in the lungs. Major adverse effects of hyperoxia on lungs include pulmonary edema, intra-alveolar hemorrhage, fibrosis, interstitial thickening, and pulmonary atelectasis [11, 12]. Hyperoxic condition also affects the central nervous system which may manifest as a condition called tonic-clonic seizure. The symptoms of oxygen toxicity induced seizure are facial twitching, vomiting, nausea, tachycardia, and disturbance in the vision. The patients with a low seizure threshold or reduced seizure threshold due to various medical conditions such as low glucose and high fever are more susceptible to hyperoxia induced seizures [13, 14]. HBOT protocols are designed to reduce the risks of oxygen toxicity along with appropriate accessory treatment and constant monitoring.

12.3 General Contraindications to HBOT

There are certain general considerations for the risk assessment for administration of HBOT with respect to the clinical condition of patients. The contraindications may be absolute or relative depending upon the intensity of their potential risks [15].

12.3.1 Absolute Contraindications

Absolute contraindications of HBOT include repercussions of pulmonary disorders and effect of oxygen toxicity. Unvented pneumothorax and acute severe bronchospasm are two such conditions which are considered as risk factors during the administration of HBOT. During the decompression phase of HBOT, untreated pneumothorax can turn into tension pneumothorax. Tube insertion is carried out for venting before the procedure to minimize the risk of iatrogenic pneumothorax [16]. Chest radiograph is used to ensure the correct positioning of tube in order to vent cavities where gas could be trapped. Acute severe bronchospasm may arise during the procedure of HBOT as a result of entrapped intrapulmonary gases which expands during the decompression phase. Patients with acute severe bronchospasm need to be provided with bronchodilators and steroids to avoid any complications during administration of HBOT [17]. Similarly, risk of oxygen toxicity and insufficiency of HBOT equipment and staff are also considered as absolute contraindications to HBOT.

12.3.2 Relative Contraindications

Relative contraindications to HBOT are based upon the potential risks with limited intensity and can be managed with relative ease [15]. Barotrauma is one of the common risk factors in HBOT use which can be managed by informing and educating the patients about autoinsufflation techniques to reduce the risks. Similarly, the incidences of oxygen induced convulsions and seizures can be avoided by keeping the administered barometric pressure below seizure threshold. Ocular oxygen toxicity and effect of hyperoxic conditions on cardiac functions need to be managed by providing frequent air breaks, use of anti-oxidants, and constant monitoring of transcutaneous oxygen tension.

12.4 Specific Cases for Contraindications

Specific physiological or pathological status of patients needs to be considered in the use of HBOT to avoid any complications. Special measures are required for the treatment of infants and young children since their respiratory capacity is not fully developed as compared to adults. Protective measures in the form of air breaks, use of anti-oxidants, and constant monitoring are employed to manage the treatment by HBOT in such cases [18]. Similarly, older patients are at higher risk of barotrauma and oxygen induced physiological disturbances due to their weakened respiratory and cardiac functions. Proper monitoring and air breaks help to manage their treatment in a safe manner. Pregnant women may experience complications during the procedure of HBOT due to the adverse effects of high pressure and hyperoxic conditions on fetus. Protocols are specifically designed in such cases to ensure the administration of HBOT in safe conditions [19]. Several pathological conditions like malignancy, diabetes mellitus, patients under specific medications, patients with medical implants, and claustrophobic patients require specific management strategies for safe HBOT administration.

References

1. Mathieu D. Handbook on hyperbaric medicine, vol. 27. Cham: Springer; 2006.
2. Mathieu D, Marroni A, Kot J. Tenth European consensus conference on hyperbaric medicine: recommendations for accepted and non-accepted clinical indications and practice of hyperbaric oxygen treatment. Diving Hyperb Med. 2017;47(1):24.
3. Karahatay S, et al. Middle ear barotrauma with hyperbaric oxygen therapy: incidence and the predictive value of the nine-step inflation/deflation test and otoscopy. Ear Nose Throat J. 2008;87(12):684–8.
4. Vahidova D, et al. Does the slow compression technique of hyperbaric oxygen therapy decrease the incidence of middle-ear barotrauma? J Laryngol Otol. 2006;120(6):446–9.
5. Rivalland G, Mitchell SJ, Van Schalkwyk JM. Pulmonary barotrauma and cerebral arterial gas embolism during hyperbaric oxygen therapy. Aviat Space Environ Med. 2010;81(9):888–90.
6. Brenna CT, et al. The role of routine pulmonary imaging before hyperbaric oxygen treatment. Diving Hyperb Med. 2022;52(3):197–207.
7. Dey D, et al. The rationale of carrying out specific investigations prior to hyperbaric oxygen therapy. Indian J Aerosp Med. 2013;57(1):56–65.
8. Workman T. Hyperbaric oxygen therapy: a high-risk procedure? Undersea Hyperb Med. 2006;33(3):222.
9. Jain KK. Oxygen toxicity. In: Textbook of hyperbaric medicine. Springer; 2017. p. 49–60.
10. McMonnies CW. Hyperbaric oxygen therapy and the possibility of ocular complications or contraindications. Clin Exp Optom. 2015;98(2):122–5.
11. Weaver LK, Churchill S. Pulmonary edema associated with hyperbaric oxygen therapy. Chest. 2001;120(4):1407–9.
12. Obiagwu C, et al. Acute pulmonary edema secondary to hyperbaric oxygen therapy. Oxf Med Case Reports. 2015;2015(2):183–4.
13. Doherty MJ, Hampson NB. Partial seizure provoked by hyperbaric oxygen therapy: possible mechanisms and implications. Epilepsia. 2005;46(6):974–6.
14. Yıldız Ş, et al. Seizure incidence in 80,000 patient treatments with hyperbaric oxygen. Aviat Space Environ Med. 2004;75(11):992–4.
15. Gawdi R, Cooper JS. Hyperbaric contraindications. In: StatPearls. Treasure Island, FL: StatPearls Publishing; 2022.
16. Inoue Y, et al. Tension pneumothorax during hyperbaric oxygen therapy. Masui. 2013;62(2):175–7.
17. Sadri RA, Cooper JS. Hyperbaric complications. Treasure Island, FL: StatPearls Publishing; 2017.
18. Lam G, et al. Hyperbaric oxygen therapy: exploring the clinical evidence. Adv Skin Wound Care. 2017;30(4):181–90.
19. Neuman TS, Thom SR. Physiology and medicine of hyperbaric oxygen therapy. Amsterdam: Elsevier; 2008.

MIX
Papier aus verantwortungsvollen Quellen
Paper from responsible sources
FSC® C105338

FSC
www.fsc.org

If you have any concerns about our products,
you can contact us on
ProductSafety@springernature.com

In case Publisher is established outside the EU,
the EU authorized representative is:
Springer Nature Customer Service Center GmbH
Europaplatz 3, 69115 Heidelberg, Germany

Printed by Libri Plureos GmbH
in Hamburg, Germany